TEXTILE FORMING

By
Meenakshi Rastogi

SONALI PUBLICATIONS
NEW DELHI-110002

Edition 2018

ISBN 978-81-8411-206-1

Published by

SONALI PUBLICATIONS
4228/1, Ansari Road,
New Delhi-110 002 (India)
Phone: 23266109, 23279245
Fax: 91-11-23253475
E-mail: sonalipub@yahoo.co.in

Printed at:

Sachin Printers, Delhi

Preface

"'Textile' literally means material that is made by uniting yarns or fibres together to from a mat or sheet of any unlimited length or width manufactured by weaving, knitting, felting, bonding etc. having been used for covering, warmth, personal adornment and even to display wealth, from time immemorial.

There are many fibrous structures in nature, but it is only these which can be spun in to yarns suitable for weaving or knitting which can be classified as textile fibres. In order that it may have commercial value, a textile fibre must possess certain fundamental properties. It must be readily obtainable in adequate quantities—at a price which will not make the end-product too costly. It must have sufficient strength elasticity and spinning power. The later property implies a measure of cohesion between individual fibres, which will give strength to the yarn when they are twisted together.

The spinning of fibres, no doubt has helped when there is a certain amount of surface roughness or serration and it is also promoted by fineness and uniformity in diameter. In addition to these fundamental properties, there are other which is desirable such as durability, softness absence of undesirable colour, and an affinity for days. Some fibres have few and others have many of these properties. Some of the valuable properties such as colour or softness may be latent and the object of finishing process is to develop them to the highest possible extent without anyway damaging the fibre.

Textile is act as the part of home science .Textile refers to Dressing, style, comfort as well as attraction or grooming. Dress serves two purposes: on the one hand it is a sign of civilized society and on the other hand it shows the level of aesthetics means science of beauty in art or nature. Dress is a sign of a personality. This book dwells on the ways to teach the readers or students, how to keep them up.

Adolescent girls are deeply conscious of their appearance. They are aware the clothing contributes greatly to their personal attractions. Clothing plays an important role in a girl's environment.

High school girls should be help to appreciate that possession of many items of clothing alone will not make them attractive, unless cleanliness and orderliness accompany them. Tidiness and neatness are important in good grooming and acceptance by others. They must also realize that character and desire to serve others are the foundations of personal attractiveness.

The natural interest of girls in clothing should be utilized to provide opportunities for them to observe and practice correct methods of clothing care. The standards of living in the homes of the community influence greatly the types of learning experiences to be selected in clothing.

A **textile** is a flexible material consisting of a network of natural or artificial fibres often referred to as thread or yarn. Yarn is produced by spinning raw wool fibres, linen, cotton, or other material on a spinning wheel to produce long strands known as yarn. Textiles are formed by weaving, knitting, crocheting, knotting, or pressing fibres together (felt).

The words *fabric* and *cloth* are used in textile assembly trades (such as tailoring and dressmaking) as synonyms for *textile*. However, there are subtle differences in these terms. *Textile* refers to any material made of interlacing fibres. *Fabric* refers to any material made through weaving, knitting, crocheting, or bonding. *Cloth* refers to a finished piece of fabric that can be used for a purpose such as covering a bed.

We hope that the discussion made in the book will help the readers to understand the different and unique aspects of Textile Forming in a most comprehensive manner to face the present and future challenges which are tackle in this field.

Thanks

Contents

	Preface	V
1.	Textile	1
2.	History of Fabrics	27
3.	Textile Fibre	41
4.	Type of Textile And Fabric	79
5.	Aspects of Textile	104
6.	Fields of Textile	125
7.	Clothing	145
8.	Fabric Selection	162
9.	Yarn and Fabrics Preparation	179
10.	Clothing Care and Sewing Techniques	198
11.	Cutting and Sewing	220
12.	Sewing Tools and Equipments	247
13.	Washing	287
14.	Finishing Touch	297
15.	Starching and Ironing of Clothes	311
	Bibliography	317
	Index	325

1

Textile

The term 'Textile' is a Latin word originating from the word 'texere' which means 'to weave'. Textile refers to a flexible material comprising of a network of natural or artificial fibres, known as yarn. Textiles are formed by weaving, knitting, crocheting, knotting and pressing fibres together. Textile Museum is that specialized category of museum which primarily preserves different types of textile and textile products.

A **textile** is a flexible material consisting of a network of natural or artificial fibres often referred to as thread or yarn. Yarn is produced by spinning raw wool fibres, linen, cotton, or other material on a spinning wheel to produce long strands known as yarn. Textiles are formed by weaving, knitting, crocheting, knotting, or pressing fibres together (felt).

The words *'fabric'* and *cloth* are used in textile assembly trades (such as tailoring and dressmaking) as synonyms for *textile*. However, there are subtle differences in these terms. *Textile* refers to any material made of interlacing fibres. *Fabric* refers to any material made through weaving, knitting, crocheting, or bonding.

Cloth refers to a finished piece of fabric that can be used for a purpose such as covering a bed.

History of Textile

The History of textile is almost as old as that of human civilization and as time moves on, the history of textile has further enriched itself. In the 6th and 7th century BC, the oldest recorded indication of using fibre comes with the invention of flax and wool fabric at the excavation of Swiss lake inhabitants. In India the culture of silk was introduced in 400AD, while spinning of cotton traces back to 3000BC. In China, the discovery and consequent development of sericulture and spin silk methods got initiated at 2640 BC while in Egypt the art of spinning linen and weaving developed in 3400 BC. The discovery of machines and their widespread application in processing natural fibres was a direct outcome of the industrial revolution of the 18th and 19th centuries. The discoveries of various synthetic fibres like nylon created a wider market for textile products and gradually led to the invention of new and improved sources of natural fibre. The development of transportation and communication facilities facilitated the path of transaction of localized skills and textile art among various countries.

Textile History of Various Countries: India

Indian textile enjoys a rich heritage and the origin of textiles in India traces back to the Indus Valley Civilization where people used home spun cotton for weaving their clothes. Rigveda, the earliest of the Veda contains the literary information about textiles and it refers to weaving. Ramayana and Mahabharata, the eminent Indian epics depict the existence of wide variety of fabrics in ancient India. These epics refer both to rich and stylized garment worn by the aristocrats and ordinary simple clothes worn by the common people. The contemporary Indian textile not only reflects the splendid past but also cater to the requirements of the modern times.

Japan

In 1869 the capital of Japan was shifted from Kyoto to Tokyo and from this time onwards the Nishijin weaving tradition seemed threatened with extinction. The industry again started to grow along with Japan's new capitalist economy by 1890 when the Nishijin weavers embraced and applied modern technology to their own ancient and original textile art. The textile art of Japan particularly reached an epitome of excellence by exhibiting a cultural distinction and remarkable artistic skill in the Edo and succeeding Meiji periods (1868 - 1912).

China

Chinese textiles enjoy an excellent heritage in textile sector and occupy a prominent position in the global textile market. Chinese textiles are world famous and extraordinary for their fine quality and profound symbolic meanings. Textiles in China often form an integral aspect of its heritage and symbolically reflect its tradition and culture. In China, textile is often closely associated with prosperity and involved in the process of elaborate rituals. Parents' spontaneous love for their children is most visibly reflected through the excellent clothes they provide on festive occasions to their children. These clothes are made up of expensive materials and excellent craftsmanship.

Africa

In the ancient times the most important aspect of textiles or more precisely cloth in Africa was that cloth was used as form of money. The width of cloth strip was usually standardized in each region of Africa and therefore there used to be a regular number of such standard length cloth strips required to make a woman's wrapper cloth. This would then be used to serve as the unit of value. Cloth was a convenient form of money primarily because it was used by everybody, fairly durable and easily sub dividable.

The weavers, dyers and other textile artists of Africa together makes an active contribution in creating exquisite and amazing range of textiles. African usually embody a great variety of styles. Adinkara, kente and bogolan are some of the African textiles which are becoming increasingly popular while some others like Yoruba, Ase-oke and Adire are equally beautiful but less well known.

The textile industry, with its extremely long and rich history, has had a massive impact on the world economy and the very evolution of modern society. Weaving is believed to be one of the oldest surviving crafts in the world today, the actual origins of which are thought to date back to Neolithic times 12,000 years ago. Even before that time, the same principle was used to interlace branches and twigs to form protective fences, shelters and baskets. Once the practicality of interlacing these kinds of materials was understood, further experimentation with other natural materials probably produced the first basic fabrics and clothes.

Uses

Textiles have an assortment of uses, the most common of which are for clothing and containers such as bags and baskets. In the household, they are used in carpeting, upholstered furnishings, window shades, towels, covering for tables, beds, and other flat surfaces, and in art. In the workplace, they are used in industrial and scientific processes such as filtering. Miscellaneous uses include flags, backpacks, tents, nets, cleaning devices, such as handkerchiefs; transportation devices such as balloons, kites, sails, and parachutes; and strengthening in composite materials such as fibre glass and industrial geotextiles. Textiles can be used for educational purposes. Textiles can be used as a material for children to use and explore in their classrooms as another element of learning. Children can manipulate and come up with creative uses for textiles such as collage materials, art materials and so on.

Textiles used for industrial purposes, and chosen for characteristics other than their appearance, are commonly referred to as *technical textiles.* Technical textiles include textile structures for automotive applications, medical textiles (e.g. implants), geotextiles (reinforcement of embankments), agrotextiles (textiles for crop protection), protective clothing (e.g. against heat and radiation for fire fighter clothing, against molten metals for welders, stab protection, and bullet proof vests. In all these applications stringent performance requirements must be met. Woven of threads coated with zinc oxide nanowires, laboratory fabric has been shown capable of "self-powering nanosystems" using vibrations created by everyday actions like wind or body movements.

Sources and Types

Textiles can be made from many materials. These materials come from four main sources: animal, plant, mineral, and synthetic. In the past, all textiles were made from natural fibres, including plant, animal, and mineral sources. In the 20th century, these were supplemented by artificial fibres made from petroleum.

Textiles are made in various strengths and degrees of durability, from the finest gossamer to the sturdiest canvas. The relative thickness of fibres in cloth is measured in deniers. Microfibre refers to fibres made of strands thinner than one denier.

Animal textiles

Animal textiles are commonly made from hair or fur.

Wool refers to the hair of the domestic goat or sheep, which is distinguished from other types of animal hair. That the individual strands are coated with scales and tightly crimped, and the wool as a whole is coated with an oil known as lanolin, which is waterproof and dirtproof. Woollen refers to a bulkier yarn produced from carded, non-parallel fibre, while worsted refers to a finer yarn which is spun from longer fibres which have been combed to be parallel. Wool is commonly used for warm clothing. Cashmere, the hair of the Indian Cashmere goat, and Mohair, the hair of the North African angora goat, are types of wool known for their softness.

Other animal textiles which are made from hair or fur are alpaca wool, vicuna wool, llama wool, and camel hair, generally used in the production of coats, jackets, ponchos, blankets, and other warm coverings. Angora refers to the long, thick, soft hair of the angora rabbit.

Wadmal is a coarse cloth made of wool, produced in Scandinavia, mostly 1000-1500.

Silk is an animal textile made from the fibres of the cocoon of the Chinese silkworm. This is spun into a smooth, shiny fabric prized for its sleek texture.

Plant textiles

Grass, rush, hemp, and sisal are all used in making rope. In the first two, the entire plant is used for this purpose, while in the last two, only fibres from the plant are utilized. Coir (coconut fibre) is used in making twine, and also in floormats, doormats, brushes, mattresses, floor tiles, and sacking.

Straw and bamboo are both used to make hats. Straw, a dried form of grass, is also used for stuffing, as is kapok.

Fibres from pulpwood trees, cotton, rice, hemp, and nettle are used in making paper.

Cotton, flax, jute, hemp and modal are all used in clothing. Pina (pineapple fibre) and ramie are also fibres used in clothing, generally with a blend of other fabrics such as cotton.

Acetate is used to increase the shininess of certain fabrics such as silks, velvets, and taffetas.

Seaweed is used in the production of textiles. A water-soluble fibre known as Alginate is produced and is used as a holding fibre; when the cloth is finished, the alginate is dissolved, leaving an open area

Tencel is a man-made fabric derived from wood pulp. It is often described as a man-made silk equivalent and is a tough fabric which is often blended with other fabrics - cotton for example.

Mineral Textiles

Asbestos and basalt fibre are used for vinyl tiles, sheeting, and adhesives, “transite” panels and siding, acoustical ceilings, stage curtains, and fire blankets.

Glass Fibre is used in the production of spacesuits, ironing board and mattress covers, ropes and cables, reinforcement fibre for composite materials, insect netting, flame-retardant and protective fabric, soundproof, fireproof, and insulating fibres.

Metal fibre, metal foil, and metal wire have a variety of uses, including the production of cloth-of-gold and Jewellery. Hardware cloth is a coarse weave of steel wire, used in construction.

Synthetic textiles

A variety of contemporary fabrics. From the left: evenweave cotton, velvet, printed cotton, calico, felt, satin, silk, hessian, polycotton.All synthetic textiles are used primarily in the production of clothing.

Polyester fibre is used in all types of clothing, either alone or blended with fibres such as cotton.

Aramid fibre (e.g. Twaron) is used for flame-retardant clothing, cut-protection, and armor.

Acrylic is a fibre used to imitate wools, including cashmere, and is often used in replacement of them.

Nylon is a fibre used to imitate silk; it is used in the production of pantyhose. Thicker nylon fibres are used in rope and outdoor clothing.

Spandex (trade name *Lycra*) is a polyurethane fibre that stretches easily and can be made tight-fitting without impeding movement. It is used to make activewear and swimsuits.

Olefin fibre is a fibre used in activewear, linings, and warm clothing. Olefins are hydrophobic, allowing them to dry quickly. A sintered felt of olefin fibres is sold under the trade name Tyvek.

Ingeo is a polylactide fibre blended with other fibres such as cotton and used in clothing. It is more hydrophilic than most other synthetics, allowing it to wick away perspiration.

Lurex is a metallic fibre used in clothing embellishment.

pes of Fibre

Fibres is a class of hair-like materials that are continuous filaments or are in discrete elongated pieces, similar to pieces of thread. They can be spun into filaments, thread, or rope. They can be used as a component of composite materials. They can also be matted into sheets to make products such as paper or felt. Fibres are of two types: **natural fibre** and **man made** or **synthetic fibre**.

Natural Fibres

Natural fibres include those made from plant, animal and mineral sources. Natural fibres can be classified according to their origin.

Vegetable fibres

Vegetable fibres are generally comprised mainly of cellulose: examples include cotton, jute, flax, ramie, sisal, and hemp. Cellulose fibres serve in the manufacture of paper and cloth.

COTTON- Cotton is a soft, staple fibre that grows in a form known as a boll around the seeds of the cotton plant (*Gossypium* sp.), a shrub native to tropical and subtropical regions around the world, including the America, India and Africa. The fibre most often is spun into yarn or thread and used to make a soft, breathable textile, which is the most widely used natural-fibre cloth in clothing today.

JUTE-Jute is a long, soft, shiny vegetable fibre that can be spun into coarse, strong threads. It is produced from plants in the genus *Corchorus*, family Tiliaceae.

Jute is one of the cheapest natural fibres and is second only to cotton in amount produced and variety of uses. Jute fibres are composed primarily of the plant materials cellulose (major component of plant fibre) and lignin (major components wood fibre). It is thus a ligno-cellulosic fibre that is partially a textile fibre and partially wood. It falls into the bast fibre category (fibre collected from bast or skin of the plant) along with kenaf, industrial hemp, flax (linen), ramie, etc. The industrial term for jute fibre is *raw jute*. The fibres are off-white to brown, and 1–4 meters (3–12 feet) long.

Jute fibre is often called hessian; jute fabrics are also called hessian cloth and jute sacks are called gunny bags in some European countries. The fabric made from jute is popularly known as burlap in North America.

FLAX- Flax (also known as **common flax or linseed**) (binomial name: ***Linum usitatissimum***) is a member of the genus *Linum* in the family Linaceae. It is native to the region extending from the eastern Mediterranean to India and was probably first domesticated in the Fertile Crescent. This is called as Jawas/Javas or Alashi in Marathi. Flax was extensively cultivated in ancient Egypt. (NewZealand flax is not related to flax, but was named after it as both plants are used to produce fibres.)

Flax is an erect annual plant growing to 1.2 m tall, with slender stems. The leaves are glaucous green, slender lanceolate, 20–40 mm long and 3 mm broad. The flowers are pure pale blue, 15–25 mm diameter, with five petals; they can also be bright red. The fruit is a round, dry capsule 5–9 mm diameter, containing several glossy brown seeds shaped like an apple pip, 4–7 mm long.

In addition to referring to the plant itself, "flax" may refer to the unspun fibres of the flax plant.

Uses- Flax is grown both for its seeds and for its fibres. Various parts of the plant have been used to make fabric, dye, paper, medicines, fishing nets and soap. It is also grown as an ornamental plant in gardens.

RAMIE- Ramie (*Boehmeria nivea*) is a flowering plant in the nettle family Urticaceae, native to eastern Asia. It is a herbaceous perennial growing to 1 - 2.5 m tall; the leaves are heart-shaped, 7-15 cm long and 6-12 cm broad, and white on the underside with dense small hairs - this gives it a silvery appearance; unlike nettles, the hairs do not sting. The true Ramie or China Grass also called Chinese plant or white Ramie is the Chinese cultivated plant. A second type, is known as green Ramie or Rhea and is believed to have originated in the Malay Peninsula. This type has smaller leaves which are green on the underside, and it appears to be better suited to tropical conditions.

SISAL- Sisal is an agave. *Agave sisalana* that yields a stiff fibre traditionally used in making twine and rope. (The term may refer either to the plant or the fibre, depending on context.) It is sometimes been incorrectly referred to as *sisal hemp* because hemp was for centuries a major source of fibre, so other fibres were sometimes named after it.

The plant originated in the Yucatan Peninsula of Mexico. In the 19th century, Sisal cultivation spread to Florida, the Caribbean islands and Brazil, as well as to countries in Africa, notably Tanzania and Kenya, and Asia. The first commercial plantings in Brazil were made in the late 1930's and the first Sisal fibre exports from there were made in 1948. It was not until the 1960,s that Brazilian production really accelerated and the first of many spinning mills were established. Today, Brazil is the major world producer of Sisal. There are both positive and negative environmental impacts from Sisal growing.Traditionally used for rope and twine, sisal has many uses, including paper, cloth, wall coverings and carpets.

HEMP- Hemp (from Old English *hænep*, see cannabis (etymology)) is the common name for plants of the entire genus *Cannabis*, although the term is often used to refer only to *Cannabis* strains cultivated for industrial (non-drug) use.

Industrial Hemp has many uses, including paper, textiles, biodegradable plastics, health food, and fuel. It is one of the fastest growing biomasses known, and one of the earliest domesticated plants known. It also runs parallel with the "Green Future" objectives that are becoming increasingly popular. Hemp requires little to no pesticides, no herbicides, controls erosion of the topsoil, and produces oxygen. Furthermore, Hemp can be used to replace many potentially harmful products, such as tree paper (the processing of which uses bleaches and other toxic chemicals, and contributes to deforestation), cosmetics, and plastics, most of which are petroleum-based and do not decompose easily. The major difference between the two types of plants is the appearance and the amount of–tetrahydrocannabinol (THC) secreted in a resinous mixture by epidermal hairs called Glandular Trichomes. Industrial Hemp is produced in many countries around the world.

This fibre can be further categorized into the following:

- Seed fibre: Fibres collected from seeds or seed cases. e.g. Cotton and Kapok
- Leaf fibre: Fibres collected from leaves. e.g. sisal and agave.
- Bast fibre or skin fibre: Fibres are collected from the skin or bast surrounding the stem of their respective plant. These fibres have higher tensile strength than other fibres. Therefore, these fibres are used for durable

yarn, fabric, packaging, and paper. Some examples are flax, jute, kenaf, industrial hemp, ramie, rattan, soybean fibre, and even vine fibres and banana fibres.

- Fruit fibre: Fibres are collected from the fruit of the plant, e.g. coconut (coir) fibre.
- Stalk fibre: Fibres are actually the stalks of the plant. E.g. straws of wheat, rice, barley, and other crops including bamboo and grass. Tree wood is also such a fibre.

The most used natural fibres are cotton, flax and hemp, although sisal, jute, kenaf, and coconut are also widely used.

Hemp fibres are mainly used for ropes and aerofoils because of their high suppleness and resistance within an aggressive environment. Hemp fibres are, for example, currently used as a seal within the heating and sanitary industries.

Animal fibres : Animal fibres generally comprise proteins; examples include silk, wool, angora, mohair and alpaca.

- Animal hair (wool or hairs): Fibre or wool taken from animals or hairy mammals. e.g. sheep's wool, goat hair (cashmere, mohair), alpaca hair, horse hair, etc.
- Silk fibre: Fibre collected from dried saliva of bugs or insects during the preparation of cocoons. Examples include silk.
- Avian fibre: Fibres from birds, e.g. feathers and feather fibre.

Industrial use of natural fibres

After World War II, the build-up of synthetic fibres significantly decreased the use of natural fibres. Now, with the increase of oil prices and environmental considerations, there has been a revival of natural fibre use within the textile, building, plastic and automotive industries. This interest is reinforced by the developmental perspectives on the agro-industrial market and local productions, allowing economic development and independence versus imported materials.

France remains the greatest European Hemp fibre producer with 50,000 tons yearly (EU 100,000 tons). France also produces the largest range of industrial seeds worldwide. China and Russia are also important producers, but the statistics in that field are not available.

In the industrial domain, the consortium DAIFA group SAS have reached a leading position in Europe in the automotive plastics market. They specialize in injection and thermopress plastics reinforced with natural fibres.

The use of natural fibres at the industrial level improves the environmental sustainability of the parts being constructed, especially within the automotive market. Within the building industry, the interest in natural fibres is mostly economical and technical; natural fibres allow insulation properties higher than current materials.

Synthetic Fibre

Synthetic fibres are the result of extensive research by scientists to improve upon naturally occurring animal and plant. In general, synthetic fibres are created by forcing, usually through extrusion, fibre forming materials through holes (called spinnerets) into the air, forming a thread. Before Synthetic fibres were developed, artificially manufactured fibres were made from cellulose, which comes from plants.

The first artificial fibre, known as artificial silk from 1799 onwards, became known as Viscose around 1894, and finally Rayon in 1924. A similar product known as Cellulose Acetate was discovered in 1865. Rayon and Acetate are both artificial fibres, but not truly synthetic, being made from wood. Although these artificial fibres were discovered in the mid-nineteenth century, successful modern manufacture began much later (see the dates below).

Nylon, the first synthetic fibre, made its debut in the United States as a replacement for silk, just in time for World War II rationing. Its nobel use as a material for women's stockings overshadowed more practical uses, such as a replacement for the silk in parachutes and other military uses.

Common Synthetic fibres include:

- Rayon (1910) (artificial. not synthetic)

- Acetate (1924) (artificial, not synthetic)
- Nylon (1939)
- Modacrylic (1949)
- Olefin (1949)
- Acrylic (ORLON)(1950)
- Polyester (1953)
- PLA (2002)

Speciality Synthetic fibres include:

- Vinyon (1939)
- Saran (1941)
- Spandex (1959)
- Vinalon (1939)
- Aramids (1961) - known as Nomex, Kevlar and Twaron
- PBI (Polybenzimidazole fibre) (1983)
- Sulfar (1983)
- Lyocell (1992)
- Zylon (PBO fibre)

Modern fibres that are made from older artificial materials include:

- Glass Fibre is used for:
  - o industrial, automotive, and home insulation (Fibreglass)
  - o reinforcement of composite and plastics
  - o specialty papers in battery separators and filtration
- Metallic fibre (1946) is used for:
  - o adding metallic properties to clothing for the purpose of fashion (usually made with composite plastic and metal foils)
  - o elimination and prevention of static charge build-up
  - o conducting electricity to transmit information
  - o conduction of heat

➛ **Rayon** is a manufactured regenerated cellulosic fibre. Because it is produced from naturally occurring polymers, it is neither a truly synthetic fibre nor a natural fibre; it is a semi-synthetic fibre. Rayon is known by the names *viscose rayon* and *art silk* in the textile industry. It usually has a high lustre quality giving it a bright shine. Rayon contains the chemical elements carbon, hydrogen, and oxygen.

- **Cellulose acetate**, first prepared in 1865, is the acetate ester of cellulose. Cellulose acetate is used as a film base in photography, and as a component in some adhesives; it is also used as a Synthetic fibre

Acetate fibre and Triacetate fibre

Acetate and Triacetate are mistakenly referred to as the same fibre; although they are similar, their chemical compounds differ. Triacetate is known as a generic description or primary acetate containing no hydroxyl group. Acetate fibre is known as modified or secondary acetate having a few hydroxyl groups. Triacetate fibres, although no longer produced in the United States, contain a higher ratio of acetate-to-cellulose than to acetate fibres.

- **Nylon** is a generic designation for a family of synthetic polymers known generically as polyamides and first produced on February 28, 1935 by Wallace Carothers at DuPont. Nylon is one of the most commonly used polymers.

Nylon is a thermoplastic silky material, first used commercially in a nylon-bristled toothbrush (1938), followed more famously by women's stockings ("nylons"; 1940). It is made of repeating units linked by peptide bonds (another name for amide bonds) and is frequently referred to as *polyamide* (PA). Nylon was the first commercially successful polymer and the first synthetic fibre to be made entirely from coal, water, and air. These are formed into monomers of intermediate molecular weight, which are then reacted to form long polymer chains.

Nylon was intended to be a synthetic replacement for silk and substituted for it in many different products after silk became scarce during World War II. It replaced silk in military applications such as parachutes and flak vests, and was used in many types of vehicle tires.

Nylon fibres are used in many applications, including fabrics, bridal veils, carpets, musical strings, and rope.

Solid nylon is used for mechanical parts such as machine screws, gears and other low-to medium-stress components previously cast in metal. Engineering-grade nylon is processed by extrusion, casting, and injection molding. Solid nylon is used in hair combs. Type 6/6 Nylon 101 is the most common commercial grade of nylon, and Nylon 6 is the most common commercial grade of molded nylon. Nylon is available in glass-filled variants which increase structural and impact strength and rigidity, and molybdenum sulfide-filled variants which increase lubricity.

Aramids are another type of polyamide with quite different chain structures which include aromatic groups in the main chain. Such polymers make excellent ballistic fibres.

Nylon fibre

The Federal Trade Commission's definition for Nylon Fibre: "A manufactured fibre in which the fibre forming substance is a long-chain synthetic polyamide in which less than 85% of the amide-linkages are attached directly (-CO-NH-) to two aliphatic groups."

- A synthetic thermoplastic fibre (Nylon melts/glazes easily at relatively low temperatures)
- Round, smooth, and shiny filament fibres
- Cross sections can be either
    - trilobal to imitate silk
    - multilobal to increase staple like appearance and hand
- Its most widely used structures are multifilament, monofilament, staple or tow and is available as partially drawn or as finished filaments.
- Regular nylon has a round cross section and is perfectly uniform. The filaments are generally completely transparent unless they have been delustered or solution dyed. Thus, they are microscopically recognized as glass rods.
- Molecular chains of nylon are long and straight variations but have no side chains or linkages.
    - Cold drawing can align the chains so they are oriented with the lengthwise direction and are highly crystalline.
- Nylon is related chemically to the protein fibres, silk and wool.
    - They both have similar dye sites but nylon has many fewer dye sites than wool.

➛ **A Modacrylic** is a synthetic co-polymer. Modacrylics are soft, strong, resilient, and dimensionally stable. They can be easily dyed, show good press and shape retention, and are quick to dry. They have outstanding resistance to chemicals and solvents, are not attacked by moths or mildew, and are nonallergenic. Among their uses are in apparel linings, furlike outerwear, paint-roller covers, scatter rugs, carpets, and work clothing and as hair in wigs.

Commercial production of Modacrylic fibre began in 1949 by Union Carbide Corporation in the United States . Modacrylic and acrylic fibres are similar in composition and at one time were in the same category. In 1960 the Federal Trade Commission decided to separate the two fibres and establish a category for each.

The Federal Trade Commission defines Modacrylic fibres as manufactured fibres in which the fibre-forming substance is any long-chain synthetic polymer composed of less than 85%, but at least 35% weight of acrylonitrile units except when the polymer qualifies as rubber.

Modacrylic fibres are modified acrylic fibres made from acrylonitriles, but larger amounts of other polymers are added to make the copolymers.

The Modacrylic fibres are produced by polymerizing the components, dissolving the copolymer in acetone, pumping the solution into the column of warm air (dry-spun), and stretching while it is hot.

Modacrylics are creamy or white and are produced in tow and staple form. If looked at in cross section views, they have an irregular shape. Modacrylic fibres are also produced in many different lengths, crimp levels, deniers and they can have various shrinkage potentials.

Current modacrylic fibre producers include Solutia Inc. in the U.S. and Kaneka Corporation in Japan.

Properties : A Modacrylic has properties that are similar to an Acrylic. However, modacrylics are flame retardant and do not combust. The fibres are difficult to ignite and are self-extinguish. In addition to a modacrylic's flame retardant properties it has a relatively high durability that is comparable to wool. Modacrylic fibres have a moderate resistance to abrasion and a very low tenacity.

Modacrylics are poor conductors of heat. The fabrics are soft, warm and resilient but are prone to pilling and matting. Modacrylics display high performance when it comes to appearance retention. The fibres are quite resilient and will not wrinkle. They also have great dimensional stability and high elastic recovery, which gives them the ability to hold their shape.

Care : Modacrylics are sensitive to loss of appearance due to improper care; therefore, it is important to know how to care for modacrylics. Modacrylics are resistant to acids, weak alkalis, and organic solvents. These fibres are also resistant to moths, mildew and sunlight. Modacrylic fabrics can be machine washed using warm water and tumble dried on a low setting. Modacrylic pieces can also be dry-cleaned, however, they should not be steamed and should only be tumbled on cold. Some fabrics may also be cleaned using the furrier method (a special non-immersion cleaning process). The fibres are heat sensitive and will shrink at 250 °F and will stiffen at temperatures over 300 °F.

Uses : Modacrylics are used primarily in applications where environmental resistance or flame retardancy is necessary or required. Modacrylics have the ability to combine flame retardancy with a relatively low density, meaning protective gear is not uncomfortably heavy (i.e. shirts and trousers worn by electrical linemen). The combination of flame retardancy and low density is also useful in furnishings, draperies, and outdoor fabrics.

Modacrylics are also commonly used in furlike fabrics, hairpieces, wigs and fleece type fabric. By mixing the various forms of fibres it is easy to create a realistic fur look. The fabrics can then be sheared or embossed to create a closer resemblance to fur. The heat-sensitivity of modacrylic also allows wigs and hairpiece to be curled and heat styled without damage.

Modacrylics are also used in fleece, knit-pile fabric backings, and non-woven fabrics. Other uses of modacrylics include paint rollers, industrial fabrics, stuffed toys and filters.

➛ **Olefin fibre** is a synthetic fibre made from alkenes. It is used in the manufacture of various textiles as well as clothing, upholstery, wallpaper, ropes, and vehicle interiors. Olefin is also referred to as polypropylene, polyethylene, or polyolefin. Olefin's advantages are its strength, colourfastness and comfort, stain, mildew, abrasion and sunlight resistance, and good bulk and cover.

Olefin fibres have great bulk and cover while having low specific gravity. This means "warmth without the weight." The fibres have low moisture absorption, but they can wick moisture and dry quickly. Olefin is abrasion, stain, sunlight, and chemical resistant. It does not dye well, but has the advantage of being colour fast. Since Olefin has a low melting point, textiles can be thermally bonded. The fibres have the lowest static of all manufactured fibres and a medium lustre. One of the most important properties of Olefin is its strength. It keeps its strength in wet or dry conditions and is very resilient. The fibre can be produced for strength of different properties.

Physical and Chemical structure

Physical : Olefin fibres can be multi or monofilament and staple, tow or film yarns. The fibres are colourless and round in cross section. This cross section can be modified for different end uses. The physical characteristics are a waxy feel and colourless.

Chemical : There are two types of polymers that can be used in Olefin fibres. The first, Polyethylene, is a simple linear structure with repeating units. These fibres are used mainly for ropes, twines and utility fabrics.

The second type, Polypropylene, is a three dimensional structure with a backbone of carbon atoms. Methyl groups protrude from this backbone. Stereoselective polymerization orders these methyl groups to the same spatial placement. This creates a crystalline polypropylene polymer. The fibres made with these polymers can be used in apparel, furnishing and industrial products.

➛ **Acrylic fibres** are synthetic fibres made from a polymer (Polyacrylonitrile) with an average molecular weight of 100,000, about 1900 monomer units. To be called Acrylic in the U.S, the polymer must contain at least 85% acrylonitrile monomer. Typical comonomers are vinyl acetate or methyl acrylate. The Dupont Corporation created the first acrylic fibres in 1941 and trademarked them under the name "**Orlon**".

Acrylic is lightweight, soft, and warm, with a wool-like feel. It dyes very well and has excellent colourfastness. It is resilient, retains its shape, and resists shrinkage and wrinkles. Acrylic has recently been used in clothing as a less expensive alternative to cashmere, due to the similar feeling of the materials. The disadvantages of Acrylic are that it tends to fuzz (or pill) easily and that it does not insulate the wearer as well as cashmere. Many products like fake pashmina or cashmina use this fibre to create the illusion of cashmere to the consumer.Acrylic is resistant to moths, oils, and chemicals, and is very resistant to deterioration from sunlight exposure.

➤ **Polyester** is a category of polymers which contain the ester functional group in their main chain. Although there are many polyesters, the term "polyester" as a specific material most commonly refers to Polyethylene Terephthalate (PET). Polyesters include naturally-occurring chemicals, such as in the cutin of plant cuticles, as well as synthetics such as polycarbonate and polybutyrate. Polyesters aren't biodegradable.

Polyesters may be produced in numerous forms such as sheets and three-dimensional shapes. Polyesters as thermoplastics may change shape after the application of heat. While combustible at high temperatures, polyesters tend to shrink away from flames and self-extinguish upon ignition. Polyester fibres have high tenacity and E-modulus as well as low water absorption and minimal shrinkage in comparison with other industrial fibres.

Woven polyester fabrics are used in consumer apparel and home furnishings such as bedsheets, bedspreads, curtains and draperies. Similarly, industrial polyesters are used in tyre reinforcements, ropes, fabrics for conveyor belts, safety belts, coated fabrics and plastic reinforcements with high energy absorption. Polyester fibrefills are also used to stuff pillows, comforters and cushion padding.

Polyester fabrics are claimed to have a "less natural" feel when compared to similarly-woven fabrics made from natural fibres (i.e. cotton in textile uses). However, polyester fabrics may exhibit other advantages over natural fabrics, such as improved wrinkle resistance. As a result, polyester fibres are sometimes spun together with natural fibres to produce a cloth with blended properties.

Polyesters are also used to make bottles, films, tarpaulin, canoes, liquid crystal displays, holograms, filters, dielectric film for capacitors, film insulation for wire and insulating tapes.

Liquid crystalline polyesters are among the first industrially-used liquid crystalline polymers. They are used for their mechanical properties and heat-resistance. These traits are also important in their application as an abradable seal in jet engines.

Thermosetting polyesters are used as casting materials, and Chemosetting polyester resins are used as fibreglass laminating resins and non-metallic auto-body fillers. Fibreglass-reinforced unsaturated polyesters find wide application in bodies of yachts and as body parts of cars.

Polyesters are also widely used as a finish on high-quality wood products such as guitars, pianos and vehicle / yacht interiors.Cured polyesters can be sanded and polished to a high-gloss, durable finish.

Mechanical properties

Energy absorption of chemical fibre reinforced plastics (impact, bending and tensile tests) Investigation of the practical requirements for measuring the energy absorption of composite materials, and development of a suitable method for carrying out such measurements. A number of dynamic testing methods for measuring the energy absorption of laminates are reviewed, including an impact bending test, repeated-impact tests, an impact tensile test, and a ram bending test. Also discussed are impact tests on plate laminates. Particular emphasis is placed in these studies on composites with a chemical fibre reinforcement. It is established that a relation exists between the quasi-static energy absorption of the fibres and the dynamic energy absorption of the composite. Composites with commercial polyester and polyamide fibres lead to the highest energy absorptions, in which the testing apparatus has a significant effect.

➛ **Polylactic acid** or **Polylactide (PLA)** is a biodegradable, thermoplastic, aliphatic polyester derived from renewable resources, such as corn starch (in the U.S.) or sugarcanes (in rest of the world). Although PLA has been known for more than a century, it has only been of commercial interest in recent years, in light of its biodegradability.

Chemical and Physical properties

Due to the chiral nature of lactic acid, several distinct forms of polylactide exist: poly-L-lactide (**PLLA**) is the product resulting from polymerization of L,L-lactide (also known as L-lactide). PLLA has a crystallinity of around 37%, a glass transition temperature between 50-80 °C and a melting temperature between 173-178 °C.

Polylactic acid can be processed like most thermoplastics into fibre (for example using conventional melt spinning processes) and film. The melting temperature of PLLA can be increased 40-50 °C and its heat deflection temperature can be increased from approximately 60°C to up to 190 °C by physically blending the polymer with PDLA (poly-D-lactide). PDLA and PLLA form a highly regular stereocomplex with increased crystallinity. The temperature stability is maximised when a 50:50 blend is used, but even at lower concentrations of 3-10% of PDLA, there is still a substantial improvement. In the latter case, PDLA acts as a nucleating agent, thereby increasing the crystallization rate. Biodegradation of PDLA is slower than for PLA due to the higher crystallinity of PDLA. PDLA has the useful property of being optically transparent.

➛ **Vinyon** is a synthetic fibre made from Polyvinyl Chloride. In some countries other than the United States, Vinyon fibres are referred to as **polyvinyl chloride fibres**. It can bind non-woven fibres and fabrics. It was invented in 1939.

It has the same health problems associated with chlorinated polymers. In the past, Vinyon was used as a substitute for plant-based filters in tea bags.

Vinyon fibre characteristics :

- Doesn't flame, but softens at low temperatures(55 C)
- High resistance to chemicals
- Moisture absorption is less than 0.5% and moisture regained is less than 0.1%

Major vinyon fibre uses :

- Industrial applications as a bonding agent for non-woven fabrics and products.

➛ **Saran** is the trade name for a number of polymers made from Vinylidene Chloride (especially polyvinylidene chloride or **PVDC**), along with other monomers. The main advantage of Saran film is a very low permeability to water vapour, flavour and aroma molecules, and oxygen compared to other plastics. The barrier to oxygen prevents food spoilage, and the barrier to flavour and aroma molecules helps food retain its flavour and aroma.

Saran is best known for having been used in *Saran Wrap* — a thin, clingy plastic wrap that was sold in rolls and used primarily for wrapping food. Saran Wrap is now a brand of S. C. Johnson & Son. Environmental concerns with halogenated materials have led to the removal of PVDC from the Saran Wrap product which is now composed of polyethylene. With this change, Saran Wrap no longer provides the excellent oxygen barrier and therefore the food spoilage prevention benefit of the original product formulation. In some jurisdictions, the name Saran is a registered trademark of the Dow Chemical Company, while in others, it has lost trademark status and become a generic term for these polymers.

➛ **Spandex** or **elastane** is a synthetic fibre known for its exceptional elasticity. It is stronger and more durable than rubber, its major non-synthetic competitor. It was invented in 1959 by DuPont chemist Joseph Shivers. When first introduced, it revolutionized many areas of the clothing industry.

"Spandex" is a generic name and not derived from the chemical name of the fibre, which most manufactured fibres are, but an extension of the word *expand*. "Spandex" is the preferred name in North America; elsewhere it is referred to as "elastane".

The most famous brand name associated with Spandex is **Lycra**, a trademark of Invista (formerly part of DuPont). Such is the prominence of the Lycra brand that it has become a genericised trademark in many parts of the world, used to describe any kind of Spandex. Invista discourages such use, protecting its trademark vigorously.

Spandex fibres are produced in four different ways including melt extrusion, reaction spinning, solution dry spinning, and solution wet spinning. All of these methods include the initial step of reacting monomers to produce a prepolymer. Once the prepolymer is formed it is reacted further in various ways and drawn out to produce a long fibre. The solution dry spinning method is used to produce over 90% of the world's Spandex fibres.

➛ **Vinalon** is a synthetic fibre, produced from polyvinyl alcohol using anthracite and limestone as raw materials. Vinalon was first developed by the Korean scientist Ri Sung Gi at the Takatsuki Chemical Research Institute in 1939. The fibre was largely ignored until Ri defected to North Korea in 1950. Trial production began in 1954 and in 1961 the massive February 8 Vinalon Complex was built in Hamhung. Its success and widespread usage in North Korea is often pointed to in propaganda as an example of the success of the *juche* philosophy. Hamhung remains a major production centre for Vinalon; in 1998, a vinalon factory opened up in South Pyongan.

The factory complex also produces other chemicals. Some North Korean defectors have claimed that Dr. Lee Seung-ki was involved in chemical weapons research and that the complex was used to produce them.

Vinalon, also known as Juche fibre, has become the national fibre of North Korea and is used for the majority of textiles, outstripping fibre such as cotton or Nylon, which are only produced in small amounts in North Korea. Other than clothing, Vinalon is also used for canvas shoes, ropes and quilt wadding.

Vinalon is resistant to heat and chemicals but has numerous disadvantages: it is stiff, uncomfortable, shiny, prone to shrinking and difficult to dye. It is not produced outside of North Korea.

➛ **Aramid** fibres are a class of heat-resistant and strong synthetic fibres. They are used in aerospace and military applications, for ballistic rated body armor fabric, and as an asbestos substitute. The name is a shortened form of "aromatic polyamide". They are fibres in which the chain molecules are highly oriented along the fibre axis, so the strength of the chemical bond can be exploited.

Aramids are generally prepared by the reaction between an amine group and a carboxylic acid halide group. Simple AB homopolymers may look like:

$$n\mathrm{NH_2\text{-}Ar\text{-}COCl}\ \text{'!}\ \mathrm{\text{-}(NH\text{-}Ar\text{-}CO)_n\text{-}} + n\mathrm{HCl}$$

The most well-known aramids (Nomex, Kevlar, Twaron and New Star) are AABB polymers. Nomex, New Star and Teijinconex contain predominantly the meta-linkage and are poly-*metaphenylene isophtalamide*s (MPIA). Kevlar and Twaron are both P-Phenylene Terephtalamides (PPTA), the simplest form of the AABB para-polyaramide. PPTA is a product of *p*-phenylene diamine (PPD) and terephtaloyl dichloride (TDC or TCl). Production of PPTA relies on a co-solvent with an ionic component (calcium chloride ($CaCl_2$)) to occupy the hydrogen bonds of the amide groups, and an organic component (N-methyl pyrrolidone (NMP)) to dissolve the aromatic polymer.

➛ **Polybenzimidazole** or **PBI** fibre (1983) is a synthetic fibre with an extremely high melting point that also does not ignite. Because of its exceptional thermal and chemical stability, it is often used by fire departments and space agencies.

PBI is prepared by step-growth polymerization from tetra-aminobiphenyl and diphenyl isophthalate spun via a dry spinning process using dimethyl acetamide as the solvent.

PBI fibre characteristics : The chemical formula of Poly[2,2'-(m-phenylen)-5,5' bibenzimidazol] (PBI) is believed to be: $([\mathrm{NH\text{-}C{=}CH\text{-}C{=}CH\text{-}CH{=}C\text{-}N{=}C\text{-}}]_2\text{-}[\mathrm{C{=}CH\text{-}C{=}CH\text{-}CH{=}CH\text{-}}])_n$ OR $(C_{20}N_4H_{12})_n$ of Molar mass 308.33608 ± 0.01764 g/mol.

- dyeable to dark shades with basic dyes following caustic pre-treatment
- resistant to most chemicals

➛ **(SULFUR) Polyphenylene sulfide** ("PPS") is an organic [polymer] consisting of aromatic rings linked with sulfides. Synthetic fibre and textiles derived from this polymer are known to resist chemical and thermal attack. PPS is used to make filter fabric for coal boilers, papermaking felts, electrical insulation, specialty membranes, gaskets, and packings. PPS is the precursor to a conducting polymer of the semi-flexible rod polymer family. The PPS, which is otherwise insulating, can be converted to the semiconducting form by oxidation or use of dopants.

Fibre Characteristics **:** PPS is one of the most important high temperature polymers because it exhibits a number of desirable properties. These properties include resistance to heat, acids and alkalies, and to mildew,to bleaches, aging, sunlight, and abrasion. It absorbs only small amounts of solvents and resists dyeing.

➛ **Lyocell** is a fibre made from wood pulp cellulose. It was first manufactured in 1987 by Courtaulds Fibres UK at their pilot plant S25. The only current manufacturer in the United States is Lenzing Inc, who market it under the trademarked brand name **Tencel**. There is another larger plant with over twice as much capacity at Grimsby in the United Kingdom.

The Federal Trade Commission defines Lyocell as "a cellulose fabric that is obtained by an organic solvent spinning process". It classifies the fibre as a sub-category of rayon.

The fibre is used in the production of many clothes, such as jeans, trousers and coats.

Major properties

Lyocell was introduced to consumers in 1991 and originally marketed as a type of Rayon. The only current manufacturer in the United States is Lenzing Inc, who market it under the trademarked brand name Tencel. Lyocell is produced from wood pulp cellulose. Lyocell shares many properties with other cellulosic fibres such as cotton, linen, ramie and rayon. Some main characteristics of Lyocell fibres are that it is soft, absorbent, very strong when wet or dry, and resistant to wrinkles; it can be machine or hand-washed or drycleaned, it drapes well, and it can be dyed in many colours, as well as simulating a variety of textures like suede, leather, or silk. FibreSource

Production method : Lyocell is created through a process called solvent spinning. The wood pulp is dissolved in N-Methylmorpholine N-oxide, creating a solution called "dope." The dope is then pushed through a spinneret to form the individual fibres. After the dope has been spun into Lyocell fibres, it is washed and the chemicals are retrieved from the water, purified, and recycled.

➞ **Zylon** is a trademarked name for a range of thermoset polyurethane synthetic polymer materials manufactured by the Toyobò Corporation. Zylon was invented and developed by SRI International in the 1980s. Like Kevlar, Zylon is used in a number of applications that require very high strength with excellent thermal stability. Tennis racquets, table tennis blades, various medical applications, and some of the Martian rovers are some of the more well known instances.

IUPAC name: poly(p-phenylene-2,6-benzobisoxazole)

2

History of Fabrics

In music, craft arclitecture and fine arts India possessed a great—heritage. India's textile product is were in great dem-and over world India. In india the textiles produced were very fme and rich with beautiful artistic designs, harmonies and good colour schemes. Silks, brocades, and muslins in artistic creations were the stamp of individuality of craftsmen who produced it. Some of the textiles varieties from deverse parts of india are as under—

1. Muslins: In producing finest grade Dacca was the expertise. The value of muslins is estimated by number of warp threads in a given length as compared with its weight. 'Malmal Khas', the finest muslin was known to have 1000 to 1800 threads in warp. The weaver took five months to weave 1.0 yards of fabric. The weaving of muslin would be done only in rainy season due to its fineness. Some of the nuslins are named as ab-i-ravan, Baft-hawa, and shabnam.

(a) *Dacca Sarees*: Till the beginning of 19th century Dacca saris were considered as most artistic speciman handlooms of Bengal women. They were gray, white, blue and black designs with silver and gold threads. Now they are available in all shades. The weaving pattern in flowers is called 'Tandani' The pallows are heavy as parders and rest of the saree is covered with bootie-s'. The flowers (boota) when spread allover is called' Bootedar I sari', when in diagonal lines is known as 'terchha' when there is network of flowers is known as jatar. When flowers are dustered like jewels, it is called , Panna hazari' .Other motifs used on pallows are 'mayura' 'hansa " animals and human figures.

(b) *Chanderi Sarees* : Chanderi is a place near Gwalior. Chanderi sarees are woven on pallavs and borders with gold threads. When silk and cotton threads are brought into use, it is known as 'Garbh reshmi' .The ground saree is checked with booties in each square, while pallows are worked with gold threads. The length is usually 9 yards which is much valued by Maharashtrian ladies.

2. Silks : Silk's cultivation is an ancient art which had its beginning in china. The empress si-ling-chi reared silk worms and spun silk thread and made a robe to present it to her husband. Chinese kept this knowledge to themselves for 3000 yrs. Then in 289 A. D. Japan and India received this knowledge. India exported much of silk textiles to western countries.

3. Baluchari : Near Murshidabad Baluchari is a town. In Baluchari saris the pallows are most exaggerate like Dacca saris with floral, bird and human figure designs. The fancied designs are Kalka, Torang, Guldasta etc. In olden times the human figures were dressed in Persian dress or European dress holding flowers or a wine glass. Still among Hindus the saris were famous.

4. Brocades : The brocades are the most gorgeous and highly ornamental of all Indian textiles. It is inter-weaving of silk threads and gold wires in beautiful colours and floral designs. It yields the most gorgeous, the most fascinating of India's silk fabrics. The designs are produced by warp and weft threads of different colours and materials suitable even today. Brocades are mainly produced in Aurangabad, Ahmedabad, Bhopal, Surat, Delhi, Banaras, Lucknow and Madras. But Banaras is their real home. A large range of textiles in Brocades are mainly done in western India. The various types of Brocades are Amru, Himru, Tanchoi, Surat

Brocades and Jarnnagar Brocades. The Amrus of Hyderabad are pure silk brocades while the Himrus are brocaded silk with a mixture of wool or cotton. Gold wire was lavishly used in ancient days to work and delicate patterns on the kimkhabs which utilised to be manufactured chiefly in Banaras. Chief motifs seen in the design are bunch .of flowers, flying birds, and water wales. The heavier the brocade the more expensive it is considered.

The designs and colours used are varied. The three chief patterns are- 1. Beldar 2. Butidar 3. Shikar Ghar.

In Banarasi kimkhab, the famous hunting scene design and sun is the fines of them all. Some of the favourite motifs are the sacred geese and animal pattern. The Kalka or the flame design is also popul. The most traditional pattern is *Beldar* .

(a) ***Himru*** : These are the shawls done in Maharashtra. Him means snow. This fabric is utilised in winter. These shawls have an over all pattern.

Here warp and weft is silk and Buti is gold.,No extra thread is seen at the back. Bigger shawls are used for dresses also. The Himru of Hyderabad is popular. Himru is an inferior kind of Brocade in which both silk and cotton threads are used to produce the several coloured designs. Today the Himrus are chiefly utilised for ladies blouses and cholies. Shervanies for man are also as furnishing fabric.

(b) ***Amru*** : Amru is commonly woven in Ahmedabad, Surat and Banaras: Silk thread is widely used for weaving.

(c) ***Tanchoi*** : These are chiefly done in western India. In tanchoi the and weft is in silk. Buties are in silk and gold both. Sometimes even silk'thread is seen. Tanchoi has no wasted lnaterial at the back that means warp and weft inserted in that makes the fabric. Heavier fabric is considered to be more beautiful. Tanchoi takes a lot of time while weaving. In a day one can weave just about 3-4" of tanchoi. Tanchoi designs are mainly parrots, birds, flowers, peacocks. Most of tanchoi have very heavy borders with rich colour and designs. Chifly blue colour is used, buties are alone in silver and gold. Silver thread is very explicitly seen. Some of tanchoi sarees are just the combination of colour without any print on it. In olden periods some of tanchoi had very heavy effect of silver and gold so that one could hardly see any silk on it. Another kind of tanchoi is done in western India. For that, chief centre is Jarnnagar. This tanchoi was called Ghaghra tanchoi. These are looked only on border and is meant only for skirt and at times daint creepers along with tiny flowers are seen above border and below border . Different kinds of brocades are utilised in large quantities in and outside Banaras.

5. Orissa -Ikat : Ikat is an Indonesian word. It is the ancient technique of creating textiles in India. India is the home for Ikat. The so known as Ikat technique is another sort of tie-dye. It is the technique reverse to Bandhan in bandhani first resist method is applied and then the fabric is designed while in Ikat, yarns are tied and dyed before being woven, that is, in the making of these clothes the warp and the weft thread are dyed separately by the tie and dye process before weaving.

That are made in many parts of India. Moreover, the well-knownpatola marriage saree of Gujarat, notable instances are the cotton shawls of Orissa and telia rumals. The chief centres of Ikat are Surat, Gujarat, Rajkot, Pochampalli and Chiriala. Majority of the that are exported to Africa. There are three types of that—

1. Single Ikat-It has only warp tied and dyed.
2. Weft Ikat-It has only weft tied and dyed.
3. Double Ikat-In this both warp and weft are tied and dyed.

In chiriala a special sort of Rumals is utilised by Muslims and the fishermen. Most of these rumals are exported to Africa and are now named as Asian rumals. In Pochampali chiefly sarees are dyed.

These sarees have geometrical designs and are done on cotton and silk both utilising single weft and double weft. These sarees are famous as wedding Sarees. The weavers who produce wedding sarees are called selirec. Their main characteristic is thai any thing made by them done in cloth has stories. Half the Asian rumals would enact mythological stories. Just one piece would form the whole story or one piece of Mahabharat. The figures done on these rumals are not at all balanced. But the cotton contrast produces the colourful effect. In recent past it is done in Hyderabad. Here these are done particular on cotton lungies. But still Gujarat and Pochampalli are the only well-known centres of Ikat. Rajkot is the Ikat training centre to follow the technical way of weaving. Since the art of Ikat was beginning to die sometimes ago and Ikat is supposed to be one of the most conventional art pieces, but now it has revived again.

6. Kanthas of Bengal : This is a folk art of Bengal. It was almost a lost art unit a few years back when it started to be revived. The particular significance of the kanthas quilt is not only in its artistic elegance but its vivid narration of tales of heroism and legends. It is a domestic art of the use of waste material. It is a treasure preserved in every home. This embroidery is done by all classes of people and worked by women only.

Kanthas were made both in East and West Bengal. They were mainly done in Jessore, Faridpur, Khulna and Mymensing all in East Bengal. Now in Bangladesh the fabrics urilised for Kanthas are made even" today by women for their personal use. The fabric utilised for Kanthas are old sarees, dhoties, placed greatly togeilier in layers in running stitches using white thread covering the whole field. The discarded old sarees are placed on top of each other. The edges are folded and tucked together. The field is filled in with fine quilt work by means of a white thread. The craftsmen embroider according to the imagination giving a pictorical account of ancient folklore in each Kantha piece.

There are five types of Kanthas. They are as given below:

1. ***Lep*** : This is a thick quilted wrap used in winter as a cover. To make it thick, a big number of saree layers are arranged one on top of the other. Designs on lep are geometrical and worked with the running stitch.

2. ***Sarlni*** : This is also a quilted wrap or cover ard used for ceremonial purposes.

3. ***Voyton*** : It is utilised as wraps for valuables etc. In shape it is square. It has a central motif and 2 or 3 borders. The size is about 3 feet square. It has wide borders of human and animal figures, a lotus in the centTel the comers being covered by trees,

flowers or leaves. This Kanthas is meant to be carried during journies or to be presented to relatives.

4. ***Oar*** : It is rectangular in shape about 2 feet by 1/2 feet in size. It is utilised for pillow cases. The designs used on this are tTees and birds or a linear pattern. The embroidery is simple but decorative, probably only a series of broad lines of leaves and tTees. Sometimes an extra boarder made by decorative stich is sewn around the four sides.

5. ***Arsilata*** : It is utilised as a wrap for mirrors and combs. It has a wide border and in the centTe lotuses, creepers or tTees are used for motif. The women in Bengal generally wear white sarees. Hence the base material was always white for Konthas. Today silks generally Kanchipuram, tussore or raw silk could be utilised as a base if they are to be discarded. The threads generally used are taken out from the borders of the discarded sarees which are utilised for the base. The colours of the threads used are blue, green, yellow, red and black. Today instead of thread drawn from the border vividly coloured silk of bright embroidery stands are used.

Stitches used are darning, satin and chair. For border stem stitches are used most commonly and typical stitch being used are very short darning stitches providing dotted line. Even a large area may be covered with closely made lines of these stitches. Back stitches are also used for continuous lines. The outline of the design is commonly in dark blue filled with stitches of same colour or in contrasting colour. Sometimes the outlines and the areas enclosed by the outline are in blue, stitches are in red. After the design has been embroidered any area left are filled with stitches in white remaining parallel to the outline of the motifs.

Konthas is utilised during festivals and rituals where the housewife ants to fulfil her vows. In each case, one has special religious designs like Mandala and Kalasa. Aplique work is also seen on the Konthas. Their two kind are found-one is utilised on flag and canopies where a notice or a lion motif is cutout in red cloth and then stitched on the white cloth. In the other type coloured cloth is cut into parochial stripes and sewn round the outlines of the design. This is generally done on pillow-cases.

The common ritual designs of Konthas are lotus, the bull with the lingo on his back, the tiger representing Goddess Durga, the swastik and the mouse reminiscent of Ganesh. The symbols, faiths, rituals and myths are often kept alive in folk arts. These are particularly and only prepared by women and depicted as folk embroideries. Kanthas still continue to be made in Bengal by women in rural areas. They make them for their personal use only.

7. Kashmir Embroidery : Kashmir embroidery and shawls are well-known for the beauty of colours, design and technique allover the world.

The best embroidery is revealed in the shawls. It is a cottage industry run by many members ofa family in their homes mainly around Shrinagar. Kashmir embroidery is practised by men and it is inevitably a commercial craft. According to Kamala S. Dongerkery, "the shawl industry in Kashmir was introduced by Zain-ul-Abidin, the ruler in the 15th century {1423-1474 A.D.). It seem that when he was yet a Prince, in Samarkand diverse crafts men worked for manufacturing different things. Zain-ul-Abidin got interested in the work of the craftsmen and when he returned to his own country he sent people in search of talented craftsmen. These shawl weavers from persia were brought to India to encourage this art.

Kind of woven Kashmir shawls :

(i) ***Pashmina shawls*** : The most of the woollen fabrics of Kashmir particularly the superior quality shawls are the Pashmina. Pashmina shawls were made from the wool of the upraharcus a species of the Asian mountain goat. Thus the shawls were known as Pashmina.

(ii) ***Dushala*** : Dushala or the double shawl are two resemble shawls stitched together so that when draped over the shoulder the wrong side was not visible.

(iii) ***Doroapha*** : This is double sided work in which there is no wrongside. Occasionally the same design is reproduced in two separate colours on the two sides creating double sided patterns. Thus it has simpler patterns with outlines and details woven in bright colours and being later on worked by hand.

(iv) ***Kasala shawl*** : These are square in shape and perhaps produced on European demand. They are commonly of twill weave or damask patterns woven them in a plain colour.

(v) ***Jamewar shawls*** : Woven completely of wool or with some cotton mixed but the floral design and brocaded parts are generally in silk or pashmina wool.

The colours most commonly used for Kashnlir shawls are white, green, purple, blue, yellow and black. They also use crimson colours (Gulnar). The motifs are mainly taken from nature. Animal and human figures are not seen in Kashmir embroidery probably due to of muslim influence in the area. The bird motifs used are parrots. magpie, wood-pecker and king fisher. The floral motifs are iris, lotus, lily, tulip and saffron flowers. The other designs are grapes, plums, cherries, almonds. The chenar, leaves is considered important motif.

The fabric utilised for embroidery is of several kind silk, cotton, wool. The threads utilised are cotton, wool and silk. The stitches utilised are simple the chief. being the satin-stitch, stem-stitch and chain-stich. Sometimes the darning stitch and the herringbone is also utilised. Kashida is the common term for Kashmir embroidery which comprises other stitches called as Zalakdozi (Chain-stitch.)

8. Doria (open work) and Talaibar (Gold work) : All the embroiders in Kashmir are men. They are trained from a very young age. Old man can do the finest work. Boys at the age of 6-7 do the outline on coarse material.

Kashmir also produces what is known as Gabhas. Gabha is a unique kind of floor covering. It is very cheap. It is devised from old or torn woollen blankets or torn shawls after being washed, milled and dyed in diverse colours. Here applique work in gay colours and bold floral or ornamental design is done. There are three types of Gabhas namely: I. embroidered, 2. applique, and 3. both embroidered and applique combined. In the applique kind of Gabhas piece of dyed blanket clothes are joined together and interlaced with embroidery.

Hook or chain stitch rugs are also very fancied of the rich designs. It done in bright as well as light colours. The material utilised for hook rugs is hessian cloth. The embroidery is done completely with thick woollen threads.

9. Namda : This is an potent and popular part of Kashmir embroidery .It is a special work on a carpet called crewel for whicli chain stitch is the base foundation supplemented by other stitches like satin, cross, filling etc. Today Kashmir embroidery is of fine quality. Many workers are seen embroidering on shawls as well as silk sarees, dress materials, table linen, cushion cover, bed cover, purses and articles of personal uses.

10. Chicken Kari-work : Chicken Kar;-Chicken Kari work ofU .P .was originated in Jammu and Kashmir.

This embroidery was first come into practice in Bengal and Dacca and later on brought to Jammu and Kashmir and Delhi. It is now seen in Bhopal, Gaya, Calcutta, Allahabad and Varanasi. But Jammu and Kashmir is a great modern centre where work of rather remarkable beauty and distinction is carried on.

1. ***Chicken Kari work*** is known as white embroidery .The fabric utilised for this work is plane white fabric. It is mostly done on white fine muslin cloth but today it is done on cambric and correspoding to fabrics. The patterns are printed from wood blocks in a washable colour. The designs are first prepared on paper and wooden stamping blocks are prepared from them. The designs are then stamped on the cloth and passed on to the embroiders. There are two kind of chicken kari work: the flat style and the knitted embossed, e.g. like jals or netting in types of design. The stitches are named as khatawa, bukhia. murri, taipchi, phanda andjali. Taipchi is a simple darning stitch used in a cheaper work. It comprise of strokes and straight lines drawn on the cloth in the direction needed to form the design. Taipchi is a flat style of chicken work.

2. ***Khatawa*** : This is an applique work prepared on white calico. This embroidery belongs to the flat style of chicken kari.

3. ***Bukhia*** : This is constituted of an invested satin stitch with designs outlined on the right side of the fabric. The thread is mainly below the cloth. Compact masson of thread are then worked on the wrong side so that the Qesign appears opaque. It is also known as shadow work. The stitches that cover the wrong side of the cloth are in herringbone producing a shadow effect.

4. ***Murri*** : It implies rice shaped. It generally done on muslin cloth.

Murri is done on the centre of the flowers. Stitch is the french knot.

5. ***Phanda*** : This correspoding to grains like millet. This is a smaller and shorter form of Murri stitch. These are utilised to fill petals and leaves in a pattern.

6. ***Jali*** : This is done by breaking up the fabric into holes and not by drawing out the threads. The warp and weft threads are pushed apart with the needles into holes and tightened to give the cloth the appearance of a net. This is done by a kind of very fine button holing and then pulling the thread aside.

Moreover this work of Jammu and Kashmir is also famous for the Gold and Silver embroidery .This embroidery is done on velvette, satin and chiffon. Sari borders, kurtas, collars, hanky and white cape are all embroidered in chiken kari of grains and now as far as 30 designs exist in chiken kari work utilising ratine, stem and back stitch, herring bone and button hole stitches.

11. Phulkari of Punjab : Phulkari ofPunjab belongs to Punjab. Phulkari literally implies flower- craft. The term phulkari is commonly utilised in East Punjab and the term Bag, for similar shawl in west Punjab, Peshawar and Rawalpindi.

Fabric used: The beauty of the Phulkari is relied on the great deal on colour of the ground material. Khaddar cloth was incessantly utilised for embroidering bagh and phulkaries. It was hand-spun and hand-woven cotton : material. The colour of the khaddar fabric was mainly red, white and blue or black. Khaddar was previously woven in narrow width. Hence two or more joints had to be made on the wrong side with running stitches different from the other. The work is very neat and is done without tracing and drawing and is divided as a work of art.

The motifs utilised are human and animal figures, floral and symbols which cover the surfaces from the four comers and reach out towards the main piece. The central design is generally a lotus with a variable number of petals and the field is interspersed with yielding patterns. The most recurrent being the tree of life, traditional animal figures, beads, chariots, kalkas and mythological stories. The borders comprise creepers, floral serials, spirals and several linear.

Thread used : It is pure silk. It is untwisted silken floss known as 'pot'. green, Golden yellow, white, crimson red and orange are the five colour prepared in selecting silk floss for phulkari work.

Motifs used : Motifs are made of horizontal, vertical and diagonal stitches yielding geometrical pattern in phulkari design while the bagh has an over all geometrically floral pattern as the name itself implies.

Stitches used : Although the darning stitch and geometrical pattern go hand and hand in phulkari still a number of designs can be yielded. The stitch of long and short darning stitches. It is a ubiquitons method of embroidery in the sense that it is worked on every side of the cloth.

The centre of this interesting art is Rohtak and the surrounding region. The best phulkari were produced in the district of Jhelum, Peshawar, Rawalpindi, Nazara and they are all now in Pakistan. In India in Punjab the phulkari embroidery was done in Amritsar, Ambala, Jallander, Patiala, Ludhiana, Rohtak and Kapurthala.

Particular authority has drawn a distinction in phulkari and bagh. In phulkari the ornamentation is dispersed mostly with large area of flower art. In Bagh the whole field is covered with pot and silk floss and not even a thread of base fabric is with beautifully mixed colours. The designs on phulkari are more or less floral in character with some geometrical pattern. Thus bagh or phulkari is the common name of these shawls with the distinction that in bagh or garden the whole surface is decorated with a connected pattern whereas in phulkari the pattern is dispersed at intervals over the earth.

Types of Phulkari : There are various types of phulkari. The motifs and designs are distirict from place to place. They may be grouped into 4 or 5 in classes. The chope and suber were wedding phulkari and were presented to the bride by the maternal relation during marriage ceremony. The plain red or dark red khaddar cloth known as saloo, were utilised in daily household ware. Til-patra shawl have very little embroidery and are of inferior quality khaddar. It was conventional custom to give servants this as presents during marriage and on auspicious occasion. Till-patra literally implies dotted with .till-seed design.

Nilak is worked on black or navy blue khaddar with yellow and crimson red pot. The nilak is pretty phulkari shawl and looks very attractive.

Kinds of Bagh : Bagh shawls are usually embroidered in one chief colour with other contrasting shape in little proportions. But two colour combinations like orange and yellow, green and crimson or white and gold look the best. One of the most splendid is the ghungat bagh or the veil shawl. It has a triangular patch of embroidery on that portion of which covers the head when worn. In this shawl there are usually two large motifs, triangular in shape and also very decorative.

Varida Bagh : The bridegroom's mother presents it to the bride. Reshmi sheesha meaning silken mirror of fine allover embroidery in white so perfect that it looks like a mirror. Bagh has several other varieties known by different names such as kakribagh, shalimar bagh, mircha bagh, dhania bagh each name derived from the motif suggested by objects with which the embroiders are familiar. The phulkari shawl would occasionally take even 10 years to be completed. Originally the phulkari embroidery covered almost the whole base and looked like the fabric itself. But the modern trend is to use the pulkari 's stitch just to embroider any cloth. Pulkari is done with great care which is a time and energy-consuming product and so each piece becomes a real heirloom handled down from generation to generation.

12. Kasuti : It is famous in the Kanataka areas. Kanataka is now in Karnataka state, which was ealier in the State of Bombay. The art of Kasuti embroidery is mainly done by women and men in house. The five garments on which kasuti was performed were kunchi (Bonnet) and cape combined; Lehga (skirt), sharagu (Pallow ofa saree), cusupa (Bodice) and Kulai (Bonnet). These were the garments the women and their children utilised and so after their children used.

A saree is generally worn with an embroidered Pallu by the Kannada women. Kulai, which is baby cap has a cape attached to the neck of the cap and also women's cholies have Kasuti design of the finest Texture.

The material on which the Kasuti embroidery was done previous was mostly khalls used as blouse pieces and Irkal sarees. Irkal is near Bejapur and the sarees produced there were generally yards in length with 3/4 of a yard as Pallow and borders about 6-8" with a varied colour in warp and weft. The pulak is in stripes of red and white and is called Teop and Tenni. Kasuti was done above the red and white stripes.

Kasuti Embroidery : It is done on any kind of fabric. It is made on cushion covers and several other household articles or hand-woven cloth.

The motifs used in Kasuti are from temple architecture. The Gopurams of South India and also lotus flowers, Palanquins, Raths, bird motifs such as parrot and peacock, snake and the squirrel are common. Animal motifs used for kasuti embroidery are the sacred bull, elephant and deer. One will rarely see mouse, lion or tiger but cats and dogs are never seen. The stitches utilised are the simplest and so minute that the effect of the work is intricate and pretty. Four kinds of stitches are used in kasuti, named such:

1. Gavanti, 2. Murgi, 3. Negi, 4. Menphi.

Kasuti is done by counting the threads. The four stitches are explained separately, as follows—

1. ***Gavanti*** : Gavanti is a line and back stitch or double running stitch. The patterns here are mainly geometrical. Since these are worked in vertical, horizontal and diagonal directions, this is the most common stitch. Here the design seems the same on both sides of the fabric.

2. ***Murgi*** : Murgi appears like the steps of ladder. As the stitches are zig-zag running stitches. It is similar to Gavanti since both are neat with the design on the wrong and right side looking alike. The stitches are regular and uniform in length and so the distance between the two stitches remains the same.

3. ***Negi*** : Negi is ordinary running of darning stitch. It has the over all effect of a woven design. This stitch is used for larger designs by weaving the stitch in conformity with the surface to be covered with thread.

4. ***Menphi*** : It has a heavy appearance and a large amount of threads needed. Hence this stitch is not of much use. Nowadays several of the kasuti workers are taking to the use of simple stitch. It is most commonly utilised to cover up the background areas of design.

13. Patola : In weave it is in plain. Before weaving the yarns are finished and dyed. The process of patola is very labourious and complex. Thus solely few designs are utilised in Patola. The designs of devnagari alphabets and mantras are woven on dhoties ofmen. Other designs utilised are Nari-Iunyar-bhat-danany girl, elephant and parrot, Pan bhat (Leaf design), Ratan Chowk bhat (Cross of diamonds), Okhas bhat (water crest), Phulvadi bhat (flowers), Chhabri bhat (basket), wagh-kunjar bhat (tiger-elephant), chawkadi bhat (Three flowers in a square).

14. Bandhani : Bhandhari or Choonaries are the odhnis dyed by tie and dye method utilised by Sindhi ladies and Rajasthanis. The fabric is folded many times and tied in varied ways and dyed to produce the pattern. The patterns rely on the method of tieing. The commonly used patterns are' Ek bundi' (one dot), char bundi (four dots), sat bundi (seven dots) and Lahariya method. The motifs used are-flowers, birds, animals, dolls. When gold work in squares is made, it is known as Gharchola.

15. Kalmedar : These are the land painted cotton fabrics. The designs are made with fine stem brush (kalam). The process is like balik. In pali pink and fabric the material is dyed is designed with wax. Ultimately the cloth is dyed in darker shaes (syah red).

3

Textile Fibre

Fibre is very resemble to hair in appearance of which clothes are made, e.g. denim, cotton, rayon, silk, nylon and polyester. The fibre is the fundamental unit of which fabric is made. When a yam is unravelled we can see a fibre. Each of the tiny hair that makes up the yam are fibres. Many fibres are twisted together into a strand. The fibre can be divided into two relying on their length :

1. *Filament fibres* : Filament fibres may be natural or man-made.

They are measured in yards or metres. Single or filaments are grouped, together to make the yam. Yam made of a single strong, smooth strand is known as mono-filamant yam, while the yam made of several small filaments is called multi-filament yam. Silk and all man-made fibres come in this class.

2. *Staple fibres* : Staple fibres are short in length and are measured in inches. Their length may range from 1-18 inches. Cotton and all small-cut fibres come in this class.

General Properties of Fibres

1. ***Hydroscopic nature***: It is the capacity of moisture absorbency of fibre. All natural fibres are hygroscopic, which also makes them stronger when they are wet.

2. ***Dimensional Stability*** : It is the stability of the structure which probihets the fibres from $hrinking because of mechanical action, agitation, triction, pressure, heat and moisture.

3. ***Elaticity*** : It is the pliability of the fibre which supports it bear the strain while they wrap around each other during twisting. This helps the fibre to oppose crushing and spring back to the original state .

4. ***Texture*** : It comprises smoothness or roughness of the fibre.

5. ***Uniformity***: Unifonn dimension supports the fibre to make better and smoother yam.

6. ***Staple*** : The dimension and length of the fibre is known as staple. The length of the fibre is related to the spinning quality , firmness, smoothness, lusture, uniformity, draping quality and strength of the yam and fabric.

7. ***Thermal property*** : It is the heat resistancy of fibre, conductivity and melting point on which the fibre melts.

8. ***Chemical property*** : It is the effect of acids, alkalies, bleaches and normal laundering reagents on the fibre.

9. ***Density*** : It is the of the molecular structure of the fibres which provides them weight.

10. ***Strength*** : The strength of the fibre makes it bear the stress and strain of spinning. The strength of the fibre is relied on the atmospheric moisture.

11. ***Resiliency*** : It is the presence of strong hydrogen bonds which prohibits the fibres from wrinkling easily.

12. ***Colour fastness*** : It is the capacity of molecular structure to hold dye molecules to retain fabric colours and brightness in sunlight, moisture and laundering.

Classification of Textile Fibres

Fibres can be divided into two classes: I. Natural Fibres, 2. Man made Fibres.

Natural Textile Fibres

These fibres grow in nature and can be divided into three groups :

(a) ***Animal fibres:*** The animal fibres which are very fanuel are wool and silk. Wool is a protective covering of camel, sheep, goat, rabbit, cat, horses and cattles. Silk is a product of silk wonn, received from its cocoon:'

(b) ***Vegetable fibres:*** Cotton and linen are the most fancied vegetable fibres. They coine from the cotton plant, a small bush resembling hollyhock. Linen is obtained from inside of wood stalk of the flax plant. These are called vegetable fibres because they comprise big amount of cellulose, of which the cells of plants are constructed.

(c) ***Mineral fibres*** : Mineral fibres are obtained from rocks. Asbestos is the most fancied mineral fibre and is utilised in fire-proof suits, protective clothing, gloves, and safety curtains in theatres.

The fibre should possess the qualities as several as feasible. There is no ideal fibre which possesses all the qualities. Some of the potent fibres and their general qualities are mentioned below—

Cotton: It is cheap, abundant and absorbant. It dyes smoothly and does not bum or scorch rapidly, it can be laundered and ironed comfartably. It wrinkles easily and becomes yellow with time: It is not affected by moth or mildew.

Silk: It is rich in look. It has natural lustre and is expensive. It can be dyed readily. It is to be laundered and ironed attentive. It can be attacked by motll or mildew. It burns and scorches smoothly.

Wool: It is a warm fibre. It is rough, durable and elastic. It is wrinkle-resistant but shrinks comfortably when moist. Laundering of wool is and bleaching too. It bums steadily and is affected by moth and mildew.

Synthetic: They can acquire the physical properties of almost all fibres. They cannot be dyed comportably and are not much absorbent. They can be mixed with other fibres. They are wrinkle-resistant and can be permanent pleated. They do not shrink and are not affected by moth and mildew.

Cotton

Origin : The cotton fibre comes from the fruit of the cotton plant which grows in tropical regions. It has lushy growth in clayey black soil and humid climate.

The cotton fibre is known as "seed hair" since it is the fibrous fluffy material which comes from the seeds of the plant. The fluffy material covering the seed is also called "ball."

Manufacture of Cotton

1. ***Ginning:*** After cotton has been picked it must be cleaned in preparation for spinning. When first picked it is called seed cotton. Only about a third of this is fibre, the rest is seed and has other uses. The seed cotton is fed into the Gin, which comprises a series of circular saws which project through steel ribs. The saws separate the lint or fibre from the seeds. It is then carried to the bailing press. There they are compressed and packaged into the standard bail of 500 lbs.

Since the first ginning does not remove all the fibre from the seeds, they are put through a second ginning process which removes the short fibres called linters. These linters are utilised in the manufacture of rayon, plastics, paper etc.

2. ***Picking:*** At the mill they are opened and fed into a mixture which arranges the fibres for uniformity .They are then processed through the picker which removes the heavier impurities such as seeds and dirt. The lint is then formed into laps which is similar to absorbent cotton.

3. ***Carding:*** The lap from the picker goes to the carding machine. This straightens the mass of fibres and lays them in parallel rows by drawing them over a revolving cylinder with teeth. The fibres emerge as a wide thin web which is then collected together into a rope or strand called a sliver.

4. ***Combing:*** Some fabrics are made from carded yarns but if a higher quality is desired, the fibre goes from the carder to the combing machine. The fibre is combed by machine until all short linters are eliminated. It is then formed into combed slivers.

5. ***Drawing:*** From the carding or combing machines the slivers then pass through the drawing operation. Many slivers are combined and drawn out into thinner strands. As a result the fibres remain parallel.

6. ***Roving:*** The twisting process starts. The sliver is drawn out into a thinner strand. and is given a few twists. The cotton is now called a rovIng.

7. ***Spinning:*** The spinning frame takes the roving and continues the process of drawing out and twisting until the finished yam emerges and is wound on the bobbins. Warp (lengthwise yarns) generally need more twisting than filling (weft or widthwise yarns) since they are subjected to more tension in the weaving process. The yam manufacturing steps discussed to this point produce a single yam.

8. ***Weaving:*** Weaving cotton yam is manly the same as weaving any other fibre. It comprises of interlacing the lengthwise warp yarns with the cross-wise or filling yarns. In the case of cotton, the warp yarns which are subjected to significant tension, may first be coated with starch or sizing to prohibit breaking or injury .Cotton yarns may be knitted as well as woven.

Cotton Fibre Morphology

Macro-structure of cotton: under the microscope cotton fibre looks as a very fine, regular fibre. It ranges in length from 3/4 to 2". The longer the fibre, the easier it is to spin into a smoother stronger yam. The fibre diameter is from .0005 to .0009 jnch. Such fine fibres allow the manufacture of fine light weight cotton fabrics, inspite of the fact that it is a dense fibre. Length to breadth ratio ranges from 6000 : 1 to 350 : I. Cotton fibres vary in colour from near white to light tan. The cotton fibre grown at sea sides are best in lustre and length.

Microscopic appearance of cotton: Under the microscope it appearts like a flat, twisted ribbon. The chief part of the fibre is rather regular, with a thick wall, a canal along the centre of the fibre known as the lumen, and about 60 convolutions per centimetre.

Cotton Fibre: (a) Longitudinal view and (b) Cross-sectional view.

Like all plant cells, cotton has a different cuticle, well developed primary and secondary walls, and a lumen. The secondary cell wall forms the bulk of the fibre and is composed of spiralling fibrils. The strength and stability of the fibre is attributed to these increasing fibrils.

Physical Properties

1. ***Tenacity*** : The strength of cotton fibre is attributed to the good alignment of its long polyroers i.e. its polymer system is about 70% crystalline, due to the countless continuouns hydrogen bond formations between adjacent polymers, and the spiralling fibrils in the primary and secondary cell walls. It is one of the few fibres which gains strength when wet. This occurs dut to the improved alignment of polymers and increase in hydrogen bond numbers.

2. ***Elasticity*** : Relatively it is elastic dut to its crystalline polymer system and for this cause cotton textiles wrinkle and crease readily.

3. ***Hydroscopic nature:*** The cotton fibre is because of absorbent, owing to the countless polar OH groups. In its polymers, these attract water molecules which are also polar. The hydroscopic nature ordinarily prohibits cotton textile materials from developing static electricity. The polarity of the water molecules. attracted to the hydroxyl groups on the polymers distribute any static change which might develop.

4. ***Thermal Properties:*** Cotton is not thermoplastic and hence excessive application of heat energy reasons the cotton fibre to char and bum, without prior melting.

5. ***Lustre*** : Lintreated cotton has no pronounced lustre. Therefore in order to make it lustrous they need to be mercerized.

Chemical Properties

1. ***Effect of Alkalies:*** Cotton fibres are resistant to alkalies and are comparatively unaffected by normal laundering. The resistance is because of the lack of attraction between the cotton polymers and alkalies.

2. ***Effect of Acids:*** Cotton fibres are weakened and destroyed by acids.

Acids hydrolyse the cotton polymer at the glycosidic oxygen atom which connects the two glucose units to form the cellobiose unit. Mineral acids being stronger than organic acids, will hydrolyse the cotton polymer more quikcly.

3. ***Effect of Bleaches:*** The most common bleaches used on cotton textile materials are sodium hypochlorite and sodium perborate. They ar-: oxidising bleaches and bleach because of the oxygen liberated from them.

4. ***Effect of sunlight and weather:*** The ultra-violet rays of sunlight provide photo chemical energy whilst the infra-red rays provide heat energy essential to degrade the cotton polymers in the pressure of atmospheric oxygen, moisture and air pollutants. The breakdown of polymers takes place through diverse hydrolysis reactions. The beginninig degradation is noticed as a slight fibre discolouration. Fading of coloured cotton textiles is partially because of the breakdown of the dye molecules in the fibre's polymer system.

5. ***Colour Fastness:*** Cotton is easy to dye and print. The classes of dye which may be used to colour cotton are ozoic, direct, reactive, sulphur and vat dyes. The polar polymer system easily attracts any polar dye molecules into the polar system. Therefore, dye molecules which can be dispersed in water will be absorbed by the polymer system of cotton.

However, the dye molecules can enter solely the amorphous regions of the polymer system of cotton. The small inter polymer spaces in the crystalline regions of the polymer system prohibit the entry of the crystalline molecules.

6. ***Mildew:*** Cotton is damaged by fungi. Heat and dampness support the growth of mildew. The fungi produces a chemical compound which has the power of changing cellulose to glucose. The fungi feed on the molecules of Sugar: Cotton treated with acrylo nitrite is resistant to mildew.

7. ***Insects:*** Moths and beetles do not change cotton. Silver fish will eat cotton cellulose especially if heavily starched.

Significance to consumer

1. ***Comfort*** : Cotton conducts moisture away from the body and permits the cooler temperature outside to reach the body, so it is cool for summers. Since short cotton fibres nap readily, cotton can be made warm when necessary. Knitted cotton underwear absorbs perspiration and keeps the wearer comfortable.

2. ***Value*** : Fabrics of cotton are inexpensive due to the fibre is relatively cheap.

3. ***Durability*** : Cotton fibre dut to its natural twist, spins well and it can be twisted very tightly. Due to this cotton yarns are strong and fabrics made from them are durable. Cottons are stronger when wet than when dry and therefore do not break on washing. Cotton can be pressed with a hot iron. Although they crease easily they do not need special particular care in pressing.

4. ***Resistance to shrinkage:*** Earlier shrinkage of cotton was a problem. Now shrinkage control treatments can be applied to cottons so that not more than 1% residual shrinkage takes place.

5. ***Ease of cleaning/washing:*** Cotton is resistant to fading by perspiration. They are resistant to alkali and hence can be washed readily with soap. They are easy to iron as damp cotton is pliable.

6. ***Versatility*** : Fabrics of several separate weights and several different constructions can be made from cotton. Cotton is proper for wearing apparel, home furnishings, industrial uses and military supplies.

7. ***Improvements in Finishes:*** Cotton can be treated for resistance to stains, water, flame, mildew and germs. It can be given special [mishes for permanent stiffness, crease resistance, crease retention, wash and wear, durable press etc.

Linen

Origin : Flax is a bast fibre taken from the stalk of the plant linen usilatissimum. Fibre processed into yarn or fabric is known as linen.

The flax plant need a temperate climate with cloudy sky and sub-stantial moisture. Flax is generally raised for one or two purposes, either for fibres or for seed. The flax plants raised for seed produces mediocre fibre and those raised for fibre do not produce good seed. Fibre flax is raised mainly in Europe with Russia producing the huge amount and Belgium the finest grade. It ranges from 12-20 inches in length.

Production of Linen

1. ***Rippling:*** The seeds and balls are stripped from the pulled stalks is known as rippling. The stalks are spread out fanwise and are pulled through the teeth of the iron comb or rippler.
2. ***Retting*** : This includes soothing the flax stalks in water to ret away the weedy core and to dissolve the gums so the fibre can be, loosened and pulled away.

   The retting process can be completed by anyone of the following procedures :

   (a) ***Dew retting*** : The flax is spread on the ground and divulged to the dew and sunlight. This takes 4-6 weeks but gives the most durable linen.

   (b) ***Pool and tank retting*** : It needs the flax to be soaked with water. The bacteria that develops in the water rets away the stalks.

   (c) ***Chemical Retting*** : After the flax is stacked in tanks, chemicals such as NaOH, Na2 CO3 or dil. H2 SO4 are added, Retting in this case if done with care, is completed in a few hours.
3. ***Scutching*** : After the flax is retted and dried, it is sent to the mill where scutching rollers break up and separate the woody portions of the stalk from the usable fibre.
4. ***Hackling:*** The next step is known as hackling. It includes cleaning and straightening of fibres by the use of combs and is basically the same as the carding and combing process in cotton manufacture.
5. ***Carding, drawing, roving, spinning, weaving:*** The fibre goes through the same processes utilised in preparing cotton yarn for the weaving operations. It comprises of further cleaning, drawing out and doubling the fibre into strong yam for warps and filling.

6. *Finishing:* Linen needs very little finishing. Bailing and bleaching are again mainly the same as in the cotton process. Some linen fabrics are also beetled. This includes pounding the cloth with large wooden blocks to give it the permanent sheen.

Fibre Properties

(1) Physical Properties :

1. ***Length:*** Flax is a long fibre running from 6 to 40" inches but averages 15-25" inches. Due to its length it is not essential to spin tbe fibres, altbough tbey can be spun lightly to hold tbe ends in.

Flax is a very strong fibre due to its very crystalline polymer system allows its very long polymers to form more hydrogen bonds than cotton polymers. It is one of the few fibres which gains strengtb when wet. This is due to the temporary improvements in polymer allignment in the amorphous regions of the polymer system. The improved allignment brings about an increased hydrogen bonds and consequent tenacity.

2. ***Elastic Plastic Nature:*** The inelastic nature of flax is because of its great crystalline structure. Its polymers get locked into position by the hydrogen bonds which are formed between the polymers. Hence the polymers cannot yield and resist displacement. This crystalline polymer system gives the flax stiffness. Flax creases and wrinkles easily since when folded tbe polymer tends to break.

3. ***Hydroscopic Nature:*** Flax is very absorbent because of innumerable polar OH groups in its polymers. These attract water molecules which are also polar. The water can only enter the polymer system in its amorphous regions, as the inter polymer spaces in the crystalline regions are too small for tbe water molecules. The hydroscopic nature prohibits the development of static electricity.

4. ***Thermal Properties:*** Flax with its very long polymers has the best heat resistance and conductivity of all the commonly used textile fibres. It is not thermoplastic and burns without melting.

(2) Chemical Properties :

Same as cotton except: Linen textile materials are not mercerised and therefore normal laundering will result in alkaline hydrolysis of the wastes and gums bonding the cells. This results in cell ends projecting above the surface of the material and is referred to as cottonising of linen. Severe cottonising weakens the material.

Importance to consumer :

1. ***Wearing Apparel:*** Flax has comparatively high strength and can be made into yarn and sheer fabrics that are strong and cool. Because of the wicking property oflip,en it is comfortable to wear. It is preferred for apparel since it is completely washable and can be boiled without harm.

2. ***Household Fabrics:*** Linen is preferred for table coverings since it wears well. It appears attractive and elegant and the yarns lie flat if a beetling finish has been applied. Beetling gives the yam a gloss too. The linen fabrics may be smooth and sheer, coarse or homespun.

3. ***Linen Thread:*** Linen thread is utilised for stitching air craft and railway carriage. Upholstery, carpets, suitcases, and materials like footballs etc.

Silk

Origin : Silk had its origin in China for more than 4000 years ago. It was a properly guarded secret in China. Several centuries passed before its knowledge has transferred into Japan via Korea. India also was the seat of silk and sericulture. Silk is the filament a silkworm spins for its cocoon. The silkworm is really the caterpillar of the silk moth. Bombyx mori, and its cocoon is the shell it constructs to protect itself during its growth from caterpillar to chrysalis to moth. A single cocoon is made of a incessant filament of silk which the silkworm extrudes from its body and throws about itself, layer upon layer, making a thick, smooth symmetrical wall.

Along with the silk filament the silkworm exits a gummy substance known as sericin. This is the cement that glues the silk filament into a firm, hard cocoon.

Sericulture

It is the culture of silk, which takes care of the caterpillar that produces the silk filament from egg through the cocoon (and with breeders to, the moth as well). It is a delicate process, requiring scrupulous care and tireless patience. The breeder moths are selected with great attention. The eggs undergo several tests to ensure perfect disease-free worms.

In cold storage they are put till the mulberry trees start to leaf. They are incubated and hatched in a week or so, into tiny silkworms which are kept under clean conditions on trays which must regularly be refilled with mulbeny leaves. The worms have insatiable appetites. The silk worms eat continuously and night and day for five weeks.

Having eaten its fill, it is placed on a pile of straw and it strats spinning its cocoon. It attaches to a twig, then moving its head from side to side it spins the filaments of silk in an endless series which build up wall within wall, caught up by the gulnmy sericin that dries and hardens on ex.posure to air. Generally the worm inside the cocoon would develop into a chrysalis and the chrysalis into a moth. The moth. would then burst the cocoon and break the one long filament into short ones. Therefore the worm inside the cocoon is devasted by stiffening it with heat before it bursts the cocoon.

Processing

1. *Reeling* (Unwinding the cocoon) : Each cocoon produces about 300 to 1600 yards of filament. The process of unwinding the filament is called reeling, and is performed in a reeling factory called filature. To yield a uniform strands of raw silk for commercial use, the filaments of 5 to 10 cocoons must be united into a single thread.

This is done by soaking the cocoons in hot water, brushing them to fmd the ends collecting and threading them through guides so that they can be twisted into a fibre of uniform strength. This is the raw silk which when wound into skeins is ready for commercial purpose.

2. *Degumming* : In this stage if silk is to be dyed it must first be degumrned. This process removes the sericin by boiling it off. An olive oil soap bath is the general method for degumming. It leaves the yam white, soft and light. The more the sericin boiled off, the better the quality of filament.

3. *Spinning:* Not all silk is thrown. Some silk is spun on the cotton system. Such silk is short silk fibre from outer and inner edges of cocoons and from broken cocoons.

Before spun silk can be processed, the gum, a sericin must be eliminated. This is done by boiling the waste silk in soapy water, rinsing, drying and processing like cotton lint.

4. *Throwing* (Twisting the threads) : The raw silk fibre as it comes from the filature is too fme to withstand the regoes of wearing. It has to' be made thicker. This involves a number of operations called throwing.

The fibres are combined and pulled and twisted to hold the filaments together. Various sorts of thrown yam are available in the market.

Fibre Properties

Molecular Structure: Silk is a natural protein fibre ranging from 600-1000 yards in length. Raw Silk is made of 2 parts-Fibroin and Sericin. Fibroin is the protein part and is made up of amino acids.

From the glands of the silkworm two filaments are extruded which are together called fibroin. They are cemented together by the soluble silk gum, sericin. The polymer system is made of layers of linear polymer. The big forces of attraction between silk polymers are thought to be hydrogen bonds. It is this alignment which gives silk its tenacity.

Silk Fibre: (a) Longitudinal view and (b) Cross-sectional view.

Physical Properties

1. ***Tenacity*** : The silk filament is strong because of its linear polymers and crystalline nature (parallel to one another). This allows more hydrogen bonds to be formed in a continuous manner. When wet, silk looses strength. This is because of water molecules hydrolyzing a great number of hydrogen bonds and in the process weakening the silk polymer.

1. ***Elastic-Plastic Nature:*** Because of crystalline nature the silk polymer cannot have much movement, if silk textile is in stretched state. Stretching , ruptures a significant number of hydrogen bonds. When stretching ceases, remain in their new positions. This disorganises the polymer system of silk which is seen ***as*** a distortion and wrinkling or creasing of the silk material.

2. ***Hydroscopic nature.*** Since silk has a very crystalline polymer system, it is less absorbant than wool, however it is satisfactorily absorbant.

The greater crystallinity of silk ' s polymer system allow fewer water molecules, to enter than does the more amorphous polymer system of wool.

3. ***Thermal properties:*** Silk is very sensitive to heat. This is because of dearth of any covalent cross-links in the polymer system of silk. The existing peptide bonds, salt linkage and hvdroeen bonds ,of silk polymer system tend to break down once the temperature go beyond 10° C. In the beginning this polymer fi.agmentation will only result in surface discolouration of silk fibre.

Long exposure to heat can result in scorching. The brown to black discolouration of silk is because of formation of minute particles of carbon. Scorching makes the fibre weak. Silk will burn in the path of flame. After removal from flame it will sputter and ultimately extinguish itself. It leaves a crisp ash and provides the smell of burning hair or feathers.

Chemical Properties

1. ***Effect of acids:*** It is degraded readily by acids. This is due to there are no covalent cross-link between the silk polymers. Therefore perspiration which is acidic will cause soon breakdown of polymer system of silk. This is noticed as weakening of silk textile.

2. ***Effect of Alkalies:*** The alkaline solution causes the silk filament to swell. This is because of partial separation of polymers by action of alkali. Salt linkages, hydrogen bonds and Vander wall force hold the polymer system together. All these forces are hydrolyzed by alkali which results in dissolution of silk filamented occurs in alkaline solution.

In the beginning the polymers get separated and prolonged exposure to alkali would result in peptide bond hydrolysis resulting in polymer degradation and complete destruction of silk polymers. Silk textiles become yellow and dull on laundering because of filament surface rearrangement of polymers as well as some polymer degradation. These two factors affect light reflection and result in yellowing or dulling.

3. ***Effect of sunlight and weather:*** The dearth of covalent cross-links in the polrmer system of silk makes it affected by sunlight. The ultraviolet rays of the sun cause peptide bonds to break which causes polymer degradation on the surface of fibres. These degradation products cause the silk fibre to not only absorb more light but to scatter the incident light to a great extent. The result is yellowing or dulling of silk.

4. ***Colour Fastness:*** Silk has obvious affinity for dyes. The lustre of silk causes its dyed and printed silk texture to appear bright.

Miscellaneous: Silk is a poor conductor of electricity which results in build use of static charges.

- Silk cali be warm depending upon the construction since it has a low heat conductivity .
- Silk is resistant to attack by mildew and is comparatively resistant to bacteria.

Silk in use: Silk has been the queen of fibres for centuries. It is used for luxury fabrics and high fashion cloths and items but itsdifabitity extends its use to sports-wear and other contemporary applications. It is preferable to dry-clean silk but certain silks can also be hand laundered. Minimum handeling and good rinsing to to make sure all soap washed off are the precautions. It should be dried in shade or ironing at medium temperature is used. Bleaching should be done only with perborate of hydrogen peroxide for chlorine will destroy the material.

Silk fibres are weakened by body peroxide and may fade colours. Many deodorants and antiperspirants contain aluminium chloride which changes silk.

Significance to consumer

1. ***Beauty*** : It is the most luxurious fabric in look. It has a natural deep lustre that makes it an aristocrat among textiles. It has a regal look. Silk fibre has a good affinity for dye and therefore the colours found in silk are numerous. It offers an incredible number in yarn structure and fabric.

2. ***Durability*** : Silk fibre is the strongest natural fibre and therefore very durable. If pure and unweighted, they will withstand for years.

3. ***Comfort:*** Silk fibre like all animal fibre is warmer than rayon, cotton linen of comparable weights. Since silk has high moisture absorbency, the fabric does not percepts damp and clammy when soaked with perspiration. For this cause blouses and underwears are made of silk.

Silk can be woven of very fine yarns in open weaves, a fact that makes silk fabrics cool in summer. Organza, georgette, net and chiffon are instances of sheer silk fabrics. Silk is light weight and therefore comfortable to carry while travelling. Also they occupy less space. Due to its elasticity, silk can be pressed into permanent pleats. Silk in dark colours don't show dirt. Silk drapes well.

4. ***Price:*** Since silk is costly, it may be a buying point for the rich.

5. ***Versatility*** : It can be utilised in sports-wear, men's and women's suits, lingerie, dress and decoration fabrics.

Difference Between Cultivated and Wild Silk

Origin : Silk is an animal fibre and is the product of two different varieties of silkworms.

Cultivated silk is produced by the mulberry silk worm, Bombyx Mori. The silk worm is domesticated and feeds on only fresh mulberry leaves. Wild silk production on the other hand is not controlled. Instead wild silk worms feed on oak leaves and produce brown, orange or green fibre. These cocoons have to be sought in the trees. The most common wild silk is that of tussar or tussah silk worn of Asia.

The medulla is a honey comb like core comprising air spaces that increase the insulating power of the fibre. The cuticle or scale layer is a warm non-fibrous Keratin. The cortex is the chief part of the fibre, made up of long finger shaped cortical cells.

Molecular Structure : Wool fibre is a protein known as Keratin. Keratin is made up of amino-acids joined by peptide linkages.

In addition to carbon, hydrogen, oxygen and nitrogen wool fibre also contains sulphur. The amino-acids present in wool are mainly bulky, while silk has chiefly small amino-acids. Due to this wool is relatively amorphous. The amino-acid cysteine present in wool ends with a thial group- SH. Two cysteine molecules can link their groups to form the more complicated amino-acid cysteine. The bond they form is known as a disulphide linkage.

H OOC-CH NH2-CH2 SH + HS CH2-NH2-CH2-COOH.
Cysteine
|
HOOC-CH-CH2-S-S-CH2-CH-COOH NH2

The disulphide bonds can form within and between the protein polymer molecules. They supports to give structure and stability to the comparatively amorphous wool fibre. Four kinds of bonding occur between molecules in the protein fibre-

(a) The weak vander wall-bonding are incessantly present.

(b) The stronger hydrogen bonding between the diverse polar side- groups on the amino-acids.

(c) The ionic bonds between the acidic and basic side-groups (salt linkages).

(d) Disulphide links which are covalent bonds.

The polymer system of wool is highly amorphous that is - 75%.

Properties of Wool

Physical properties :

1. ***Tenacity*** : Wool is a complicated weak fibre. The low tensile strength is because of comparatively fewer hydrogen bonds.

When wool absorbs moisture, the water molecules steadily force sufficient polymers apart to cause a significant number of hydrogen bonds to break. The water molecules also hydrolyse several salt linkages in the amorphous regions of the fibre. Breakage and hydrolysis of these inter-polymer forces of attraction are explicit as swelling of the fibre and result in loss of strength of the wet woollen material.

2. ***Elasticity and resiliency:*** Wool fibre is elastic and resilient. Covalent bonds can stretch, but they are strong. The disulphide bonds in the amorphous parts of the wool fibre are able to stretch when the fibre is extended. When the fibre is released the disulphide bonds pull the protein molecules back into their original positions.

If there are too few disulphide linkages as when the fibre has been weakened by alkali or if the extension is great enough to break some of the covalant bonds, then some polypeptide chains will slide past one another . This causes a permanent extension of the wool. The natural crispness of the fibre also supports it to regain its real shape.

3. ***Hydroscopic nature:*** The very absorbent nature of wool is because the polarity of the peptide group, the salt linkages and the amorphous nature of the polymer system. The peptide groups and salt linkages easily attract water molecules which enter the amorphous polymer system of the wool fibre. In comparatively dry weather wool may develop static electricity .This is since these are hot enough. Water molecules in the polymer system supports to distribute any static electricity which might develop.

4. ***Density*** : Wool has a comparatively low density and therefore wool fibres are light with regard to their visible weight.

5. ***Conductivity of heat:*** Wool has a low conductivity of heat and therefore makes it ideal for cold weather. The resiliency of the fibre is significant in the warmth properties of the fabric. Wool fibres do not pack well in yarns because of the crimp and scales, and this makes wool fabric process and capable of inserting much air. Air is one of the best insulators since it keeps body heat close to the body. The medulla of the wool fibre comprises air spaces that increase the insulating power of the fibre.

Wool fibre can take up moisture in vapour form. Absorbency is a factor also in the warmth of clothing. In wider, when people go from a dry indoor atmosphere into the damp outdoor air, the heat developed by the fibre in absorbing moisture keeps to protect their bodies from the impact of the cold atmosphere.

6. ***Dimensional stability:*** Wool has poor dimensional stability and therefore shrinks easily. Felting or shrinkage results since under mechanical action, such as agitation, friction and pressure in the presence of heat and moisture, the wool fibre tends to move root wards, and the edges of the scales interlock prohibeting the fibre from returning to its original position. This results in the fabric becoming thicker and smaller, that is it shrinks or felts.

Chemical Properties

1. ***Effect of acids:*** Concentrated acids damage wool since they hydrolyse the salt linkages and hydrogen bonds. Dilute acids do not affect wool.

2. ***Effect of alkali:*** Wool easily dissolves in alkaline solutions. Alkalies hydrolyze the disulphide bonds; hydrogen bonds and salt linkages of wool and cause the wool polymers to separate from each other, which is looked as dissolution of the wool fibres. Hydrolysis of the peptide bonds of wool polymers leading to polymer fragmentation and total destruction of the wool fibre. Prolonged exposure to alkalies causes hydrolysis of the peptide bonds of wool polymers leading to polymer fragmentation and total destruction of the wool fibre.

3. ***Effect of bleach:*** Chlorine bleach is ordinary harmful to the wool. KM_nO_4' H_20 and Na_20_2 are utilised for bleaching.

4. ***Effect of Sunlight and Weather:*** Effect of Sunlight's ultra-violet radiation tends to yellow white-or dull coloured fabrics. The ultraviolet rays cause the peptide and disulphide bonds to sever. This leads to polymer degradation products on the surface of the wool fibre. As a consequence the fibre not only absorbs more light but scatters the incident light to a greater extent. The prolonged exposure to sunlight very weakens the fibres.

5. ***Colour fastness:*** Like cotton wool is easy to dye. Ac d dyes, chrome and mordant dyes are utilised to dye wool. The dye molecules are attracted into the amorphous areas of wool.

Wood's Production wool : Sheep are sheared to remove the fleece m the spring season. Wool removed from animals that have been slaughtere is referred to as 'pulled wool' .Pulled wool is inferior in quality to fleece or tip wool as it is less lustrous and elastic. Wool can be spun into yarns by the woollen or the worsted system.

1. ***Sorting the fleece:*** Each wool fleece yields many diverse grades of wool fibres. A skilled sorter can separate them by qualities called sorts.

The shoulder wool has generally the best fibre. Fine fibres that are comparatively long are reserved for sheer wool fabrics and for worsteds. Medium fibres of shortet length are suitable for woollens while coarse fibres are suitable for carpets.

2. ***Scouring:*** It removes oils, grease, perspiration and some of the dirt and impurities from the fleece. The fleece is washed a number of times in a warm, soapy alkaline solution. Scouring converts a fleece of 8 Ibs to one of 3-4 Ibs. The weight loss results from the removal of lenolin. This lenolin is recovered for use in cosmetics etc;

3. ***Carding:*** All wool fibres are carded. Fine wire teeth mounted on a cylinder separate the fibres and make them some what, almost totally parallel. This procedure also removes remaining vegetable matter from the fibre. Yarns that have been carded are known as woollen yarns. Woolen yarns are carded only, and the fibres are rather random in arrangement. They still contain foreign matter which is removed by carbonization i.e., treating with surface acid.

4. ***Spinning:*** Spinning puts in the needed twists. The slivers are wound from one spool to another and while doing so the yam is provided a twist. These yarns are much thicker as compared to cotton and silks. The yarns are sold to weavers on these spools or in skeins.

5. ***Twisting:*** Twisting is spinning two, three or four yarns together (piles). Two ply yarns are used for weaving and machine-knitting. Three and four ply are utilised for hand-knitting.

6. ***Reeling:*** Worsted yarns are wound around and reeled into skeins.

7. ***Inspection:*** Inspecting the skeins and putting them into bundles of 40 Ibs. each completes worsted yam manufacture.

8. ***Combing:*** Before combing the decision is made whether the fibres are to be utilised for woolen or worsted yarns and fabrics. Combing aligns longer fibres in a parallel manner so that the resulting "Worsted y ams " are smoother and stronger. The fibres are combed by passing through the combing machine. Here the fibres are laid in a very oriented arrangement.

Short fibres are removed and the remaining fibres are pulled into a twisted strand known as 'top'. The short fibres are known as 'noils'. Wool tops can be dyed in this form or worsted yarns can be constructed and then dyed.

Difference between Worsted and Woolen System

1. Worsteds are chosen for special uses where it is desirable to get a hard surface finish, greater durability, and sharper colouring.
2. Under the worsted system, the treatment of wool from the raw state to the fmished state makes it possible for the fibre to be twisted in such a manner, as to create a smoother or harder surface. This is the essential difference. Fibres can be twisted before to weaving.
3. Worsted fabrics generally have smoother texture than woolens.
4. Worsted are made up of a better grade of yam.
5. Worsted do not easily sag and hand a crease well.
6. Worsted takes shine more easily due to the twist.
7. Worsteds last longer than woollens.
8. Worsted cloth is less bulky and lighter than woolens.
9. Worsted cloth has greater tensile strength than woolens.
10. Worsted are easy to tailor and drape.

Significance to Consumer

1. ***Appearance:*** Wool seems attractive partly due to its natural low lustre for good grade wools or steam lustred finish for fabrics like wool broad cloth The softness of wool and the way it is woven in worsteds suits to the consumer Wool has a rich colouring The colours can be soft or bright and permanent. It drapes and fits well.

2.. ***Ease and Care:*** The wrinkle-resistant quality of wool makes it appropriate for clothes of daily wear Worsteds can hold a rease and therefore are preferred for slacks. Wool does not easily now dirt due to its resilience.

Wool absorbs moisture gradually and therefore any liquid which soaks into it can easily be squeezed out and along with it dirt too. The scales on the fibre make it water-repellent and therefore liquid can be run off the fibre if it is slanted. Improvements in shrinkage control will lessen felting.

3. ***Hand:*** The softness and springy percept and wool is a pleasure to the consumer Woolens are made of short fibres and therefore have fuzzy appearance which makes them warmer than worsteds

4. ***Suitability*** : Wool fabrics are made in several weights of very fme to very heavy. Women's garments can be made of very fine wool Sports- clothes, men's wear, smooth lustrous fabrics are the diverse ranges ofwool- wear Worsteds are excellent for suits for both men and women. Worsteds are easily tailored and drape we Woolen blankets can be both light or heavy.

Differenee between Silk and Wool

1. ***Campasition*** : Wool has hydrogen, carbon, sulphur, nitrogen. Silk has carbon, hydrogen, oxygen and nitrogen. As a consequence wool is harmed by moths and beetles while silk is not attacked by them. However silk is attacked by carpet beetles

2. ***Elasticity*** : Wool has folded poly-peptide chains while silk has extended poly peptide chains Therefore wool is more elastic and resilient

3. ***Strength:*** Silk is very crystalline whereas wool has more amorphous areas. Silk is strong while wool is more absorbent.

4. ***Dimensional Stability*** : Silk is solid fibre while wool has four parts to fibre It has a scaly outer layer Therefore wool shrinks and felts.

5. ***Texture:*** Silk is smooth. White wool fibre has crimp which is molecular. Therefore wool is warmer and more resilient. White silk is smoother and more lustrous.

6. ***Length:*** Silk fibre is usually a filament while wool is staple fibres. For this cause wool is fuzzy while silk is smooth.

7. ***Hydroscopic nature:*** Since silk has a very crystalline polymer system it is less absorbent than wool. The greater crystallinity of silk polymer system permits fewer water molecules to enter than does the more amorphous polymer system of wool.

8. ***Thermal properties:*** Silk is more sensitive to heat than wool. This is believed to be partly due to the dearth of any covalent cross-links in the polymer system of silk compared with the disulphide bonds in wool.

9. ***Effect of acids:*** Silk is degraded more than wool. Because of absence of disulphide bonds in silk, perspiration which is acidic will cause breakdown of silk polymer.

Man-made fibres : Man-made fibres may be divided into two chief categories: (a) regenerated,and (b) synthetic. There is a tendency to call all the manufactured fibres synthetic, but a careful distinction should be drawn between the two. Regenerated fibres need natural organic material as a initial point. This raw material is usually of high molecular weight and is degenerated in the process of fibre manufacture. It can, therefore, hardly rank as synthetic. The regenerated fibres may be divided into: (a) cellulose (b) protein.

Cellulosic regenerated fibres are viscose rayon and acetate rayon. Viscose rayon is prepared from cellulose of wood pulp. Acetate rayon is made by regeneration of cellulose under tension to give a filament of high tensile strength. Protein regenerated fibres are prepared with casein, soyabean and Zean. In casein,the milk mixed in alkali is extruded into acid by coagulating baths comprising aluminium salts and formaldehyde.

Rayon Fibre

Manufacturing Process

The principal raw materials utilised for making rayon is wood pulp. But cotton linters (the cotton fibres are too short for yam manufacture) are also utilised. The wood pulp and linters, are processed and cellulose extracted from it and formed into sheets.

The sheets of cellulose are covered with alkali and permitted to soak until alkali cellulose is made.

The alkali cellulose is then shredded into alkali cellulose crumbs and aged for a specific' time. The cellulose crumbs are then treated with carbon disulphide which incessantly changes the white cellulose to orange sodium cellulose xanthate. The xanthate crumb is mixed in sodium hydroxide and forms a honey coloured liquid which is aged to reach the right viscosity. The viscous liquid is forced through a spinnerette into sulphuric acid which generates the cellulose in a continuous filament.

While the fibres are hardening they are stretched :

(c) to reduce the diameter of the fibr.

(d) to align the molecules in the fibre;

(a) to permit the fibre to stretch without breaking; and

(b) to increase tensile strength.

For filament yarn most of viscous rayon fibre is twisted into Yarn, reeled into skins, washed, bleached and dried. For two, huge bundles of mono filament yarns are accumulated without twist and cut into staple fibres to be spun like natural fibres.

Types of Rayon: I. Viscous Rayon, II. Cuprarnonium Rayon, III. Saponified Cellulose Rayon.

Properties

Molecular structure: Rayons are regenerated cellulose. Length of the molecular chain relies upon the type of natural cellulose that was utilised and on the spinning process.

Rayon Fibre*:* (a) Longitudinal view and (b) Cross-sectional view

Regular rayon has a much shorter chain length than cotton, lower crystallinity and less orientation of the molecules. Its core is very amorphous and its skin is somewhat oriented. Molecular structure and chemical composition are responsible for most of its properties.

Physical Properties

1. ***Strength:*** Rayon has varying strengths. Regular rayon looses some of its strength when wet. The highly wet molecules of rayons are strong. ,

2. ***Absorbence*** : Absorbency relies on inner structure of regenerated , cellulose, which is highly amorphous. It can be penetrated readily by water and other aqueous solutions. Due to its absorbency it can be readily dyed and it absorbs perspiration easily. As water enters the fibre, it makes the hydrogen bonds between the molecular chains, pushes them apart and causes the fibre to swell. In this swollen state the fibres are weaker. Garments made of rayon shrink easily and clothes dry steadily.

3. ***Resiliency:*** Rayon has low resiliency due to the weak hydrogen bonds. Fabrics wrinkle easily unless given a resin finish.

4. ***Thermo Chemical:*** Same as that of cotton. If temperature is too high rayon scorches and then burns.

5. ***Density*** : Same as that of cellulose.

6. ***Conductivity*** : It is enough to prevent the build up of static electric charges.

Chemical Properties

1. ***Effect of Alkali:*** Strong alkali causes rayon fibres to swell and loose strength. Weak alkalies do not destroy.

2. ***Effect of Acids:*** Hot and concentrated acids cause rayon fibre to disintegrate.

3. ***Miscellaneous:*** Ultra-violet rays of sun deteriorate rayon. They are resistant to all insects barring silver fish. Mildew destroys rayon.

Acetate Fibre

Origin: Schutzen Burger in 1869 prepared accetated fibre. It is a cellulose accetate prepared by heating cellulose with acetil anhydride and glacial acetic acid and sulphuric acid at low temperature.

Manufacture of cellulose Acetate : Cellulose is acetylated in closed mixing tanks with big stirrers. It is cooled while circulating between the temperature of35° F to 1200 F. When the fibrous structure of linters vanishes, the solution is forced through fine holes in coagulating medium. Coloured filaments can be made by mixing suitable dyes. They are changed into filaments by dry spinning in warin air. The filaments are then gethered and guided to the winds. These filaments are then twisted and packed as continuous filament acetate for wearing and knitting mills. Acetate fibre can be made of desired diameter staple, length and crimp. To make staple, fibre is cut by cutting unit and crimped.

Microscopic appearance : Under the microscope it seems as continuous and smooth. The cross-section dimlges two, three or four or more smoothly rounded webs which are distinguishable from striated edges.

Physical Properties

1. Acetate filament: It is in broken strands. The length relies on the capacity of winding mechanism. Modern winders hold many pounds of yarns with one or two knots per cone.

2. Acetate staple fibre: It can be cut into desired length. Commonly desired lengths are 1.50, 2, 2.50 and 3 inches till 7 inches for silk industries.

3. Fineness: The diameter of the fib!e ranges from 1.5 to 4.1 denier for filaments and 1.5 to 20 denier for staple fibres.

4. Appearance and colour : The lustre and colour may be as liesired. Bright lustre may be received by lesser twist. Staple are not as Justrous as filaments. Various shades from paler tints to heavy, may be acquired as they are the whitest fibres.

5. Density : The density ranges from 1.25 to 1.33 gms. per cc. relying on the methods of measurement and moisture at the time of making.

6. Absorption :The hygroscopicity is correspoding to to cellulose bufabsorption is much less since hydroxyl groups are removed by acetyl groups.

7. Tenacity : The wet strength is 65 percent of dry strength. The fibre can be changed in tenacity by the manufacturers.

8. Elasticity : Under general situations, elastic recovery of fibre is complete with in few seconds. But for longer time and heavier load creep, deterioration or clearage occurs.

Thermal properties

It is a thermoplastic material. With temperature it becomes

soft. It can be ironed at the temperature upto 250° F. Above this sticking and melting occurs.

1. Conductivity : It is an perfect insulator. The static electricity is created by friction.

2. Effect of alkalies: It is stable to water even at boil and can withstand soap solutions and alkalies at normal temperature.

3. Effect of acids: It is unaffected by thin solution but is attacked by strong acids. The degradation of molecules occurs and the chain breaks.

4. Effects of oxidizing agents: Mild oxidizing agents may be utilised i.e. chlorine is old and mild peroxide.

5. Effect of organisms: It is resistant to attack by bacteria and mildew but it is attacked by moth which make holes in fabric.

Importance to consumer *:* It is utilised for men's, women 's and children's apparel, household textiles, Industrial textiles and military textiles since it is harmless if ingested (by infants), nontoxic, non-i-,Titating to skin and non-allergenic and versatile.

Synthetic Fibres

Synthetic fibres are produced in the test-tube. The fibre forming substances are not natural bases but complex chemical compounds, which are utilised for non-textiles and coatings of textiles, e.g. nylon, acrylic, polyester.

Acrylic

Acrylic fibres are manufactured fibres in which the fibre forming substance is a long chain synthetic polymer composed of at least 85% by weight of acrylonitrtite units.

H2C-CH
I
CN

Manufacture

All acrylic fibres are made of at least 85% acrylo nitrite. The remaining portion can be of any other organic channel similar in structure with a double bond. Addition polymerisation takes place in the monomers add on one to another to form a linear chain with several molecular segments. The average number of molecules in a polymolecule of acrylic fibre is about 2000.

The structural formula of acrylic is :

```
H  H  H  H  H  H  H  H
I  I  I  I  I  I  I  I
C - C - C - C - C - C - C - C
I  I  I  I  I  I  I  I
H  CN     HCNHCN  H CN
```

Then the polymer is dissolved in dimethyl formanide or dimethyl acetanide. The dissolved polymer is extruded through and spinerettes into wann air where the fibres solidify as the solvent evaporates. Surface alkaline hydrolysis takes place steadily leading to surface discolouration, due to dulling of the fabrics.

Physical Properties

1. ***Molecular structure:*** Acrylic polymer is one of the longest synthetic fibre polymer made. It has a crystalline polymer system which is about 80% crystalline and 20% amorphous. The chief forces of attraction are the vander wall forces.

2. ***Tenacity*** : Strong tenacity due to crystalline structures well as to the very long polymers. The vander wall forces though weak are effective due to the increased alignment of the polymers. There is loss of tenacity when wet because of the fact that water molecules enter the amorphous system.

3. ***Elastic Plastic nature:*** Though the polymer system is crystalline, the acrylic polymers can slide over each other when the staple fibre or filament is bent or crushed. This displacement of the polymer system is looked as wrinkles on the fabric when crushed.

5. ***Effect of sunlight*** : They are the most sunlight and weather resistant fibres. When exposed to sunlight they will first suffer a small loss in tenacity after which there is a loss in levelling off tenacity in any further reduction in tenacity. From there on they have excellent sunlight and weather resistance.

The 'levelling off' is because of a slight internal polymer rearrangement i.e., the heat energy causes properties of some polymers to assume ring structures which have more stable electron arrangement and this is able to withstand ultra-violet radiation.

6. ***Hydroscopic nature:*** Acrylity fibres are hydrophobic because of the crystatlline polymer system. Very few water molecules are absorbed due to the slight amorphous nature. Static electricity readily develops due to the death of water molecules to dissipate it.

7. ***Thermal properties:*** Acrylic fibres are most heat sensitive of the synthetic fibres. The weak wearing forces contribute to this heat sensitivity . Body heat, and the stresses and strains of wearing can give sufficient heat energy to reduce the cohesive effectiveness of the forces. This is explicit as wrinkling. With a flame, they ignite easily and burn like nylon.

Chemical Properties

1. ***Effect of acids:*** They are resistant to acids since the polymers do not attract acid.

2. ***Effect of alkalies:*** Crystalline nature prohibits entry of alkaline substances.

Polyester

Definition : Polyester fibre is defined as a manufactured fibre in which the fibre forming substance is any long-chain synthetic polymer made of at least 85% by weight of an ester of a subaromatic carboxylic acid comprising, but not confined to dimethyl terephthalate units and parasuls hydroxy benzoate units.

Manufacture

1. ***Raw materials:*** The raw materials from which polyesters are prepored is petroleum from which the constituent acids and alchohols are derived. The most widely used acid is terephthalic acid or dimethyl terepthalate and the alcohol used is ethylene glycol.

2. ***Polymerisation*** : Condensation polymerisation takes place in the pariod of reaction of the acid and the alcohol, in a vacuum at high temperature.

```
    O     O     H    R
    I     I     I    I
O - C - C - O - C - C
                I    I
                H    H    n
```

n = 80-100

Every fibre molecule or polymer comprises about 80 of the above units. Polymerised material is derived in a kind of a ribbon through holes in a spinerette. These strands enter a cooling through where they solidify. The ribbon hardens and little polyester chips are cut from the ribbon.

3. ***Spinning:*** The chips are dried to remove any residue moisture. 11ley are then put into hopper reservoirs for melting. The melted polymer (melt-spun) is extruded through the spinerette, which solidifies on heating the air and is wound loosely onto cylinders.

4. ***Drawing:*** Before the filaments can be utilised they must be drawn and expanded to impact strength control elongation. Fibres are stretched to about five times their original length which decreases their width.

The drawn fibre is wound onto cones as filaments or crimped and cut into staple lengths.

Properties

The polymer system: The polymer system is calculated to be about 80% crystalline and 15% amorphous. Very weak hydrogen bonds exist between the polymer system. The most potent forces of attraction in the system are vander wall's forces.

For vander wall's forces to be effective, excellent polymer orientation is needed. and this occurs in the very crystalline polyester polymer system.

Physical properties

1. ***Tenacity*** : Polyester filaments and staple fibre are strong due to their crystalline nature. The crystalline nature permits for the formation of highly effective vander wall's forces as well as since hydrogen bonds which provided the fibre its good tenacity .The tenacity remains unchanged when wet since the fibre resists the entry of water molecules to a significant extent.

2. ***Elastic plastic nature:*** The very crystallinity of the fibre prevents wrinkling and creasing. Repeated stretching and straining causes, distortion of the polymer system as the vander wall's forces cannot withstand much stretching.

3. ***Hydroscopic nature:*** Polyester filaments and staple fibres are hydrophobic. The lack of polarity and the very crystalline structure resists the entry of water molecules into the polymer-system. The hydrophobic nature of the polyester system attracts, fats, greases, oils, acid or any other greasy soils. It is believed to be oleophilic. The oleophilic, nature makes it not easy to remove grease by soap but by dry-cleaning with organic solvents.

4. ***Thermal Properties:*** It is a poor heat conductor and it has low resistance to heat.

It melts on heating. Polyester textile materials can be permanently heat-set. It is a thermoplastic fibre meaning that it is capable of being shaped or turned when heated. Thermoplastic fibres heated under strictly controlled temperatures soften and can then be made to similar to a flat, creased or pleated configuration. When cooled tilermoplastic fibres retain the new configuration.

Chemical Properties

1. ***Effect of Acids:*** Polyester polymers are resistant to acids.

2. ***Effect of alkalies:*** Alkaline conditions as seen in laundering hydrolyse the ester groups in polyester polymers. The crystalline nature prohibits hydrolysis to a greater extent and it is the surface of filament which gets hydrolysed. Continued laundering results in hydrolysis and materials get fewer as the surface film of the fibre gets lost.

3. ***Effect of bleaches:*** It does not require bleaching. It retains its whiteness and requires only chlorine bleaches to be used when essential.

4. ***Sunlight:*** It withstands the sun's ultra-violet radiations and is resistant to acidic pollutants in atmosphere.

5. ***Colour Fastness:*** It is not easy for dye molecules to penetrate the fibre when dyed, it retains its colour after regular wash.

6. ***Micro-Organism:*** It is resistant to bacteria and other micro- organism.

Nylon

The word nylon does not imply to one single polymer but to a large, family of them which can be defined as manufactured fibres in which the fibre fonning substance is a long chain synthetic polyamide in which less than 85% of the amide (-C-N+) 'inkages are adhered directly to two aromatic rings.

Nylon Fibre: (a) Longitudinal view and (b) Cross-sectional view

The two most common kind of nylon in production today are- (i) nylon 6, 6, and (ii) nylon 6. The nomenclature is based on the number of carbons in the repeating units within the molecular structure. RG -nylon 6, 6 is made up of adipic acid containing 6 carbon and poly-hexa-methylene- diamide comprising 6 carbons.

Manufacture of nylon 6, 6

1. ***Chemicals for synthesis:*** The chemicals from which nylon 6,6 is synthesised are adipic acid and hexamethylene diarnide.

Adipic acid	Hexamethylene diamide
COOH $(CH2)_1$ + COOH	NH_2 $(CH2)_6$ NH_2
1+4+1=6	6

There are 6 carbon atoms in adipic acid and 6 carbon atoms in hexamethylene diamine. For this reason nylon is designated as 6, 6. From Benzene or Butadene these compounds are synthesized.

2. ***Polymerization:*** The first step in the manufacturing is to cause reacting materials to form long chain molecules or polymers. Nylcn is composes by condensation polymerization. The reaction of two chemicals takes place in the free atmosphere.

Water which is split off during polymerization is permited to escape from the reacting tank. If delustred nylon is desired, then titanium dioxide is added during step. The molten polymer that is formed is extruded from tank as a ribbon many inches in width. The material is quenched in cold water which reduces the sizes of crystals formed.

3. ***Spinning:*** The ribbon is broken in small nylon chips. Nylon is melt spun. The chips fall into electrically heated grid that is too small to permit the chips to pass through. The chips melt here. The molten nylon forms a pool of liquid that is filtered to remove any impurity .This molten nylon passes through orifices in a spinerette and filaments are made. Cold air is blown across the filament which hardens them. The fibre is then passed through a steam chamber to permit the fibre to absorb some moisture. This establishes an equilibrium so that the fibre will not change later when it comes in contact with moisture.

4. ***Cold drawing:*** In this stage nylon is not very strong or lustrous. Comes orient the molecules in the nylon fibre and thereby increase both lustre and strength, the. cooled fibre is drawn in its length.

$HOOC\ (CH_{2})_4\ COOH + NH_2\ (CH_2)6\ NH_2\ -H_20 + HOOC\ (CH_2)4\ CO\ -N\ I + (CH_2)_6\ NH_2$

hexamethylene diamine adipamide.

Physical Properties

1. ***Tenacity*** : It has very good tenacity due to its crystalline polymers system and the excellent potentials of the nylon polymers to form hydrogen bond. Hydrogen bond of the strongest ring are formed i.e. those bet hydrogen and oxygen atoms. Due to this a very durable strong and elastic fibre is formed. When wet, it looses some of its tenacity due to hydrolysis of a important number of hydrogen bonds in the amorphous regions of the polymer system.

2. ***Elastic plastic nature:*** The very good elastic property of nylon fibre is because of regularity of hydrogen bonds in its system. These hydrogen bonds prohibit polymer slippage and cause polymers to return to their original position in the polymer system, once the strain has been removed from polymer system. Therefore no wrinkles and creases are formed in nylon. However severe straining con too to breakage ofbonds slippage and therefore wrinkling. Nylon is the toughest textile fibres.

3. ***Hydroscopic nature:*** Nylon fibre is not absorbent even though there is strong attraction for water molecules by polar amide

groups. Nylon crystalline polymer system permits few water molecules to be absorbed. Nylobn material easily develops static elec-tricity as they are not able to absorb sufficient water molecules to prohibit any build up of it.

4. ***Thermal Properties:*** Nylon fibres have low heat conductivity and low heat resistance. Heat causes the nylon polymers to become excited and the result is a break down of inter polymer bonding. The bonding is restored on cooling since most of the broken hydrogen bond are restored.

The excessive heat causes the polymer to become so excited that majority of nylon material melts. More, heat will result in burning. If controlled heat is applied so as to break a few hydrogen bonds. The material can be heat- set. During heat setting of thermoplastic -material the objective is to break those hydrogen bonds which are under strain to make able the material to assume the mantioned configuration.

Chemical Properties

1. ***Effect of acids:*** To acids Nylon is less resistant. The amide groups are hydrolysed by acids. Fragmentation takes place and weakens fibres. Perspiration which is acidic or acid pollutants in the atmosphere will cause some polymer hydrolysis on the surface of polymers. This changes their light reflection resulting in yellowing of white nylon and dulling of coloured nylon.

2. ***Effect of alkali*** : Long and frequent exposure to alkali results in hydrolysis of nylon polymers, and the fibre weakens.

3. ***Effect of sunlight:*** The fibre has a fair resistance to sunlight. Polymer degradation takes place following ultra-violet rays and it causes weakening of fibres.

4. ***Effect of bleaches:*** Tensile materials are generally white and do not need bleaching. Oxidising bleaches are good.

5. ***Colour fastness:*** Acidic, disperse and premel dyes are used.

Significance to consumer

I. Durability : Nylon has exceptional tensile strength and elasticity and also lightness which makes it feasibly to be made into sheer fabrics. Nylon resists abrasion and is therefore last longer.

2. ***Suitability*** : Its resiliency and strength makes it useful for number of uses. It is suitable for underwear and hosiery .Stockings do not become baggey at knees due to nylon's resiliency.

3. ***Ease and care:*** Heat-setting of nylon makes pleats perlalltent.

Heat-setting also prevents nylon from shrinking and sagging. They are easy to wash, many of them being drip-dry. Nylon dries very quickly. Filament nylon dries faster than spun nylon. There is no need to mouth-proof nylon. It is not affected badly by water, perspiration or dry-cleaning agents but midle, may discolour it.

4. ***Versatility*** : Its versatility is limitless especially in its blends with other fibres. Nylon blended with rayon makes rayon stronger, abrasion resistant and easy to wash and wear. When nylon is mixed with cotton it enhances the strength, dimensional stability , better resistant to perspiration, better elasticity and rapid drying properties.

With acrylic fibres it improves its washing quality. Nylon staple is useful for socks, sweater, swim suits, diapers, upholstery, rugs, industrial fabrics, darning threads and lingerie. It is excellent for carpets. It is used for non-textiles which are strong and elastic.

Protein Fibres

Natural Protein fibres are of animal origin. Protein Fibres are those fibres in which the fundamental chemical structure is composed of amino-acids joined in poly peptide chain. All protein fibres comprise the elements, carbon, hydrogen, oxygen, and nitrogen. Wool comprises sulphur as well. In each protein fibre these elements are combined in diverse arrangements. As a result properties of the various protein fibres may show some striking differences. However these share so many of common properties

1. ***Resiliency:*** Protein fibres have greater resiliency than cellulosic fibres. They are more resistant to wrinkling and hold their shape.

2. ***Protein*** fibres tend to be destroyed by dry heat. Wool becomes harsh and brittle and scorches readily with dry heat. It

should be ironed as in the case of silk, with a press cloth or steam. White silk and wool turn yellow. They need lower ironing temperature than cotton and linen, with recommended temperatures around 500°F .

3. ***Strength:*** Protein fibres except silk, tend to be weaker han cellulose fibres and they are weaker when wet than dry Therefore they have to be handled attentively. During washing wool looses 40% of its strength.

4. ***Static Electricity*** : They are poor conductors and hence build up static charge, so the resulting yarns and fabrics tend to release much static electricity, althought this problem is subsided in presence of moisture.

5. ***Density and specific gravity*** : Their density and specific gravity tend to be lower than that of cellulose. Fabrics made from these fibres are felt lighter in weight than comparable fabrics made from cellulose.

6. ***Hydroscopic Nature:*** Protein fibres have excellent moisture absorbency and they absorb extra moisture at the saturation point.

7. ***Thermo Chemical Properties:*** Fibres from protein family do not burn easily. When set aflame they may extinguish themselves. Burned fibres smell like burning hair; flesh or feathers. They form black crushable ash.

Therefore they are easy in wet damp climate though they tend to be warmer than cellulose fibres. Moisture prohibits brittleness in carpets.

8. ***Alkali:*** Natural protein fibres have poor resistance to alkalies and can be dissolved in 5% solution of NaOH at the boiling point. When washing neutral or little alkaline soap or detergent should be utilised. Perspiration weakens the fibre .

9. ***Bleaching:*** Chlorine bleaches destroy fibre and therefore should not be used. Fabrics can be bleached safely with hydrogen peroxide.

10. ***Sunlight:*** Sunlight disco lours fabrics made from protein fibres, turning yellow after extended exposure. Although wool has

better resistance to sunlight than cotton, it will degrade prolonged exposure to sunlight.

Fibre Family

Protein: Peanut Protein, Corn Protein, Milk Protein, Soyabean Protein.

Identification of Fibres

The textile fibres can be identified with the support of diverse methods. The consumer at shop can identify the fibre with the use of his senses.

1. ***Sense of touch*** : It comprises warmth, coolness, pliabili~, texture bumpiness, smoothness, or, strength of the fabric. A fabric that returns to its original shape and form after stretching is believed be elastic i.e. wool. For comparison of the texture, cotton and velveteen: Silk has smooth slippery surfrace, cotton has a soft surface. The vegetable fibres are generally cooler than animal fibres. The sense of touch is not very reliable in supporting the consumer to relognise the fabrics.

2. ***Sense of sight:*** Lustre, fuzziness, flatness, fineness and, coarseness may be observed. A pure silk has rich lustry surface, cotton, flannel has jazzy surface; linen has a flat, smooth surface; silk chiffon is fine and sheer; rayon and shantung looks bumpy.

Laboratory methods of fibre identification

Fourteen varied classes of fibres (acetate, acrylic, nytril, polyester, rayon, cot ton, flax, silk and wool etc.) can be recognised by nine tests.

1. ***Resistance to heat and flame*** : As a part of preliminary inspection this test should be applied to note the burning characteristics, burning odour and effects of heat. The fibre is pushed directly into the flame. It is cooled and the ash is checked.

2. ***Miscroscopic examination:*** Fibre is examined in longitudirial and cross-sectional photo micrographs. It should be done with low magnification (50 to 60). If more microscopy is required, group of fibres should be mounted at a higher magnification

(250 to 500). The longitudiual appearance should be compared with photo micrographs of known fibres. If further examination is required cross-section-1 appearance should be studied.

3. ***Solubility test*** : Fibres are mixed or disintegrated in selected liquids such as organic solvents as well as acids and alkalies.

4. ***Stain test*** : Diverse manufacturers of dyes make particular stains for fibre identification. The tests show typical colours resulting from staining the principal fibres. The procedure is simple, but recognition may be not easy since in an intimate mixture, all fibres may be stained about the same colour.

5. ***Melting point test*** : A single fibre is placed between 19 - millimetre micro cover glasses on a calibrated melting point apparatus. The fibre temperature is raised to the melting point.

6. ***Moisture regain test*** : In this test all fibres can be separated from each other. A specimen fibre is dried, weighed and conditioned in air at 700 F., 65 percent relative humi.dity. To condition the moisture-regain, the percentage increase in recorded..

7. ***Specific gravity test*** : A specimen fibre is placed in a liquid of knowns specific gravity, observed to determine whether it sinks or floats.

8. ***Breaking test*** : Mravelled thread from selvage and weft is held between fmgers and pulled, In cotton and linen, they break. Linen is stronger than cotton. The broken ends are brushy.

9. ***Tearing test*** : In cotton and linen when fairly big piece is torn the sound produced is sharper in linen. The threads along the torn edge comprise straight and pointed while in cottop-are fluffy and brushy.

Other tests : Other tests include a refractive index test and an infrared spectrum test. Heat discolouration test is user at the temperature of 3500 F for a particular time.

4

Type of Textile and Fabric

A textile is a cloth, which is either woven by hand or machine. "Textile" has traditionally meant, "a woven fabric". The term comes from the Latin word texere, meaning to weave.

Fibres are the raw materials for all fabrics. Some fibres occur in nature as fine strands that can be twisted into yarns. These natural fibres come from plants, animals, and minerals. For most of history, people had only natural fibres to use in making cloth. But modern science has learned how to produce fibres by chemical and technical means. Today, these manufactured fibres account for more than two-thirds of the fibres processed by U.S. textile mills.

Plants provide more textile fibres than do animals or minerals. Cotton fibres produce soft, absorbent fabrics that are widely used for clothing, sheets, and towels. Fibres of the flax plant are made into linen. The strength and beauty of linen have made it a popular fabric for fine tablecloths, napkins, and handkerchiefs.

The main animal fibre used for textiles is wool. Another animal fibre, silk, produces one of the most luxurious fabrics. Sheep supply most of the wool, but members of the camel family and some goats also furnish wool. Wool provides warm, comfortable fabrics for dresses, suits, and sweaters. Silk comes from cocoons spun by silkworms. Workers unwind the cocoons to obtain long, natural filaments. Fabrics made from silk fibres have great luster and softness and can be dyed brilliant colours. Silk is especially popular for scarves and neckties.

Most manufactured fibres are made from wood pulp, cotton linters, or petrochemicals. Petrochemicals are chemicals made from crude oil and natural gas. The chief fibres manufactured from petrochemicals include nylon, polyester, acrylic, and olefin. Nylon has exceptional strength, wears well, and is easy to launder. It is popular for hosiery and other clothing and for carpeting and upholstery. Such products as conveyor belts and fire hoses are also made of nylon.

Most textiles are produced by twisting fibres into yarns and then knitting or weaving the yarns into a fabric. This method of making cloth has been used for thousands of years. But throughout most of that time, workers did the twisting, knitting, or weaving largely by hand. With today's modern machinery, textile mills can manufacture as much fabric in a few seconds as it once took workers weeks to produce by hand.

Woven fabrics are made of two sets of yarns - a lengthwise set called the Warp and a crosswise set called the Filling or Weft. The warp yarns are threaded into a loom through a series of frames called Harnesses. During the cloth-making process, the harnesses raise some warp yarns and lower others. This action creates a space, or shed, between the yarns. A device called a Shuttle carries the filling through the shed and so forms the crosswise yarns of the fabric. The pattern in which the harnesses are raised and lowered for each pass of the shuttle determines the kind of weave.

Knitted fabrics are made from a single yarn or a set of yarns. In making cloth, a knitting machine forms loops in the yarn and links them to one another by means of needles. The finished fabric consists of crosswise rows of loops, called courses, and lengthwise rows of loops, called Wales. This looped structure makes knitted fabrics more elastic than woven cloth. Garment manufacturers use knitted fabrics in producing comfortable, lightweight clothing that resists wrinkling. Other fabrics include tufted fabrics, nets and laces, braids, and felt. None of these fabrics is woven or knitted. However, the textile industry produces another class of fabrics specifically called Non-Woven Fabrics.

Cotton

Cotton is a soft white fibrous substance covering seeds of certain plants.

History of the use of cotton : Asiatic cottons first grew wild in East Africa. About 5000 years ago, the people in what is now Pakistan cultivated cotton. The army of Alexander the Great first brought cotton goods into Europe in the 300s B.C. but the cloth was too expensive and only the rich could afford it.

The English began to weave cotton in the 1600s. They imported raw cotton from other countries bordering the Mediterranean Sea. Later, they imported cotton from southern colonies in America.

In the 1700s, English textile manufacturers developed machines that made it possible to spin thread and weave cloth into large quantities. Today, the United States, Russia, China and India are major producers of cotton.

Production of cotton : When cotton arrives at a textile mill, several blenders feed cotton into cleaning machines, which mix the cotton, break it into smaller pieces and remove trash. The cotton is sucked through a pipe into picking machines. Beaters in these machines strike the cotton repeatedly to knock out dirt and separate lumps of cotton into smaller pieces.

Cotton then goes to the carding machine, where the fibres are separated. Trash and short fibres are removed. Some cotton goes through a comber that removes more short fibres and makes a stronger, more lustrous yarn.

This is followed by spinning processes which do three jobs: draft the cotton, or reduce it to smaller structures, straighten and parallel the fibres and lastly, put twist into the yarn. The yarns are then made into cloth by weaving, knitting or other processes.

After inspectors check the cloth, it is passed through a gas flame that singes the fuzz off its surface. Boiling the cloth in an alkaline solution removes natural waxes, coloured substances or discolourations. Then the cloth is bleached in hypochlorite or peroxide. The cloth may then pass through a machine that prints designs on it. Cloth intended to be solid-coloured goes through a dye bath.

For fabrics made of 100% cotton:

- Set your washing machine to the low setting.
- Wash the cloth in cold water.
- Set your iron to Medium to iron it.
- Do not bleach the cloth.
- Hang to dry but do not tumble dry.

50% Cotton and 50% Polyester:

- Turn it inside out and machine wash with warm water.
- Wash with fabrics that are of light colours.
- Use only non-chlorine bleach to bleach the cloth.
- Tumble dry with a medium setting.
- Set your iron to medium for ironing.

Properties of the cotton:

- It is soft and comfortable.
- It wrinkles easily.
- It absorbs perspiration quickly.
- It has good colour retention and is good to print on.
- Cotton is also strong and durable.

Uses : Cotton is used to make blouses, shirts, singlets, bermudas, jackets and pants. It can also be made into bedspreads, pillowcases, carpets and curtains.

Wool

Woolen fabric is made from the fleece of sheep.

History of the use of wool : Wool was probably the first animal fibre to be made into cloth. The art of spinning wool into yarn developed about 4000 B.C. and encouraged trade among the nations in the region of the Mediterranean Sea.

The first wool factory in England was established in 50 A.D. in Winchester by the Romans. In 1797, the British brought 13 Merino sheep to Australia and started the the country's Merino sheep industry.

There are 40 different breeds of sheep in the world producing a rough estimate of 200 types of wool with varying standards. The major wool producers in the world are Australia, Argentina, China and South Africa.

Production of Wool : The processing of wool involves four major steps. First comes shearing, followed by sorting and grading, making yarn and lastly, making fabric.

In most parts of the world, sheep are sheared once a year, in early spring or early summer. The best wool comes from the shoulders and sides of the sheep.

This is followed by grading and sorting, where workers remove any stained, damaged or inferior wool from each fleece and sort the rest of the wool according to the quality of the fibres. Wool fibres are judged not only on the basis of their strength but also by their fineness (diameter), length, crimp (waviness) and colour.

The wool is then scoured with detergents to remove the yolk and such impurities as sand and dust. After the wool dries, it is carded. The carding process involves passing the wool through rollers that have thin wire teeth. The teeth untangle the fibres and arrange them into a flat sheet called a web. The web is then formed into narrow ropes known as silvers.

After carding, the processes used in making yarn vary slightly, depending on the length of the fibres. Carding length fibres are used to make woolen yarn. Combing length fibres and French combing length fibres are made into worsted yarn.

Woolen yarn, which feels soft, has a fuzzy surface and is heavier than worsted. While worsted wool is lighter and highly twisted, it is also smoother, and is not as bulky, thus making it easier to carry or transport about.

Making worsted wool requires a greater number of processes, during which the fibres are arranged parallel to each other. The smoother the hard-surface worsted yarns, the smoother the wool it produces, meaning, less fuzziness. Fine worsted wool can be used in the making of athletics attire, because it is not as hot as polyester, and the weave of the fabric allows wool to absorb perspiration, allowing the body to "breathe".

Wool manufacturers knit or weave yarn into a variety of fabrics. Wool may also be dyed at various stages of the manufacturing process and undergo finishing processes to give them the desired look and feel.

The finishing of fabrics made of woolen yarn begins with fulling. This process involves wetting the fabric thoroughly with water and then passing it through the rollers. Fulling makes the fibres interlock and mat together. It shrinks the material and gives it additional strength and thickness. Worsteds go through a process called crabbing in which the fabric passes through boiling water and then cold water. This procedure strengthens the fabric.

Care of your woolen garment : It is better to hand wash your woolen products. Shampoos can be used to wash such garments. Use a good shampoo not one containing "crème rinse". If shampoo can clean your hair without leaving residues, it can clean your woolen products too. Be careful of what softener you use because softeners may add products to your wool. Some types of softeners may leave a coating on the fibre making it feel softer and smoother, but at the same time causing your wool to lose its absorbency.

Properties of the wool :

- It is hard wearing and absorbs moisture.
- It does not burn over a flame but smoulders instead.
- It is lightweight and versatile.
- Wool does not wrinkle easily.
- It is resistant to dirt and wear and tear.

Uses : Wool is used to make sweaters, dresses, coats, suits, jackets, pants and the lining of boots. It can also be made into blankets and carpets.

Silk

Silk is the fine strong soft lustrous fibre produced by silkworms.

History of the use of silk : The Chinese has used silk since the 27th century B.C.. Silk is mentioned by Aristotle and became a valuable commodity both in Greece and Rome. During the Roman Empire, silk was sold for its weight in gold. The Chinese domesticated silk worms and fed them with mulberry leaves. They unwound the silkworms' cocoons to produce long strands of silk fibre.

Farm women in China at that period were supposed to raise such silkworms as one of their chores. Silk was used in China and exported along the Silk Road (the ancient trade route linking China and the Roman Empire). This trade brought China great wealth, but the Chinese did not give away the secret on how silk was formed. Christian monks finally broke China's monopoly of the silk production by smuggling silkworm eggs out of the country, and soon other countries started to produce their own silk.

Production of silk : Silkworms are cultivated and fed with mulberry leaves. Some of these eggs are hatched by artificial means such as an incubator, and in the olden times, the people carried it close to their bodies so that it would remain warm.

Silkworms that feed on smaller, domestic tree leaves produce the finer silk, while the coarser silk is produced by silkworms that have fed on oak leaves. From the time they hatch to the time they start to spin cocoons, they are very carefully tended to. Noise is believed to affect the process, thus the cultivators try not to startle the silkworms.

Their cocoons are spun from the tops of loose straw. It will be completed in two to three days' a time. The cultivators then gather the cocoons and the chrysales are killed by heating and drying the cocoons.

In the olden days, they were packed with leaves and salt in a jar, and then burried in the ground, or else other insects might bite holes in it. Modern machines and modern methods can be used to produce silk but the old-fashioned hand-reels and looms can also produce equally beautiful silk.

Care of your silk garment : It is best to dry clean your silk garment either by individual or bulk method, in order to maintain the characteristic of the silk.

Dry clean: Sending it to the dry cleaners may be done but make sure that you inform the dry cleaner that your garment is made of silk.

Hand wash :

- Make sure you wash in cool water.
- Use a small amount of soap or mild detergent to wash the silk.
- After washing it, rinse in cold water.
- Hang it or lay it flat in a shaded area to dry.
- If pressing is needed, use an all cotton iron board cover, a low or moderate steam setting, and press on the wrong side of the fabric while ironing.

Other Tips :

- To keep white silk white, add peroxide and ammonia to the wash.
- Do not use bleach or any washing detergent with bleach.
- To remove yellow from white silk, add a few teaspoons of white vinegar to the wash.

Properties of the silk :

- It is versatile and very comfortable.
- It absorbs moisture.
- It is cool to wear in the summer yet warm to wear in winter.
- It can be easily dyed.
- It retains its shape and is relatively smooth.
- It has a poor resistance to sunlight exposure.
- It is the strongest natural fibre and is lustrous.

Uses : Silk is used to make blouses, dresses, scarves, pants and ties. It can also be made into curtains, draperies, cushion covers and sofa covers.

Nylon

Nylon is a strong, light synthetic fibre. It is a man-made.

History of the use of nylon : This synthetic fibre was introduced in the 1930s. It was another early substitute for silk and soon quickly became the fibre of choice for women's stockings.

The first commercial production of nylon in the United States was in 1939 by the E.I. du Pont de Nemours and Company. Wallace H. Carothers, a chemist of the Dupont Company played the leader in this development of nylon. This man-made fibre is the second most used fibre in the United States.

Production of nylon : Nylon is made by forcing molten nylon through very small holes in a device called a spinneret. The streams of nylon harden into filament once they come in contact with air. They are then wound onto bobbins. These fibres are drawn (stretched) after they cool.

Drawing involves unwinding the yarn or filaments and then winding it around another spool. Drawing makes the molecules in each filament fall into parallel lines. This gives the nylon fibre strength and elasticity.

After the whole drawing process, the yarn may be twisted a few turns per yard or metres as it is wound onto spools. Further treatment to it can give it a different texture or bulk.

Other Tips :

- Extensive washing and drying in an automatic dryer can lead to piling.
- White nylon should be washed separately to avoid it turning grey.
- To avoid yellowing of your white nylon, you should bleach it frequently with Sodium Perborate bleach.

Properties of the nylon :

- It is strong and elastic.
- It is easy to launder.
- It dries quickly.
- It retains its shape.
- It is resilient and responsive to heat setting.

Uses : Nylon is popular for hosiery and is used to make track pants, shorts, swimwear, active wear, windbreakers, bedspread and draperies. It can also be made into parachutes, flak vests, combat uniforms, tires and life vests, or used as the netting of bridal veils, umbrellas and luggage.

Polyester

Polyester is the general name for any group of widely used synthetic products. Polyesters are strong, tough materials that are manufactured in a variety of colours, shapes and sizes.

History of the use of polyester : The first commercial production of polyester was by the du Pont de Nemours Company. It is the most used fibre in the United States.

Production of polyester : Polyesters are made from chemical substances found mainly in petroleum. Polyesters are manufactured in three basic forms - fibres, films and plastics.

Polyester fibres are used to make fabrics. Poly (Ethylene Terephthalate or simply PET) is the most common polyester used for fibre purposes. This is the polymer used for making soft drink bottles. Recycling PET by re-melting it and extruding it as fibre saves much raw materials as well as energy.

PET is made by ethylene glycol with either terephthalic acid or its methyl ester in the presence of an antimony catalyst. In order to achieve high molecular weights needed to form useful fibres, the reaction has to be carried out at high temperature and in a vacuum.

Care of your polyester garment :

- Machine-wash your fabric in cold water.
- Wash with light colours.
- Use a gentle cycle while washing.
- Tumble dry with the setting low.
- Do not bleach.
- If needed, iron with a cool iron.

Properties of the polyester :

- It is resists wrinkling.
- It is easy to launder.
- It dries quickly.
- It is resistant to stretching and shrinking.

Uses : Polyester is used to make most forms of clothing like shirts, running shorts, track pants, windbreakers, and lingerie. It can also be made into curtains and draperies.

Textile of Different Countries

India Textile : Heritage of textile in India. In India text enjoys a rich heritage and the origin of **Textiles in India** trac back to the Indus Valley Civilization where people used homespu cotton for weaving their clothes. Rigveda, the earliest of the Vec contains the literary information about textiles and it refers t weaving. Ramayana and Mahabharata, the eminent Indian epic depict the existence of wide variety of fabrics in ancient India These epics refer both to rich and stylized garment worn by the aristocrats and ordinary simple clothes worn by the common people. The contemporary Indian textile not only reflects the splendid pa but also cater to the requirements of the modern times. The ric tradition of textile in India has been favored by a number of factors The favorable factors leading to the extensive growth of **Textile tradition in India** follows:

- Easy availability of abundant raw materials like cotton, woc silk, jute and many more.
- Widely prevalent social customs.
- Variety of distinct local culture.
- Constructive geographic and climatic conditions.

Each and every region of India contributes in creating a myriad of textile tradition. The hilly region of the country produces a rich variety of woolen textiles. The pashmina and shahtoosh shawls of Kashmir, shawls and woolen garments of Himachal Pradesh and other north eastern states provides excellent examples of world famous woolen **Indian textiles**. The barren and semi barren regions like Rajasthan and Gujarat usually prefers embroidered.

The coastal areas of the south eastern regions prefer light coloured fabrics and particularly cotton and silk textiles are very popular over here. The home furnishing utilitarian textile products like bedspreads and sheets, pillows and cushions, linens and mats, curtains and napkins, carpets and rugs and many such other items are produced by all parts of the country.

Distinct style and tradition of Indian textiles :

• Silk and cotton weaving predominates the rich tradition of weaving in India. Silk weaving is most popular in various parts of the country. Assam, Banaras, Mysore, Surat, Kanchipuram and Surat are all important centers of silk weaving. All these centers specialize mostly in cotton and silk saree weaving and some of the popular traditional Indian sarees are Banarasi, Patola, Baluchari, Pochampalli, Paithani and many others.

- The ornate Style of appliqué Most commonly done bright coloured fabrics from the Kutchh region of Gujarat is very popular in the country as well as in abroad.
- The state of Punjab is famous for its phulkari work.
- Lucknow, a city in Uttar Pradesh is famous for its chikan work

Apart from those mentioned, each and every region and state of India has its own distinct individual style in textile.

Chinese Textile

China enjoys an excellent heritage in textile sector and occupies a prominent position in the global textile market. Chinese textiles are world famous and extraordinary for their fine quality and profound symbolic meanings. Textiles in china often form an integral aspect of its heritage and symbolically reflect its tradition and culture.

In China, textile is often closely associated with prosperity and involved in the process of elaborate rituals. Parents' spontaneous love for their children is most visibly reflected through the excellent clothes they provide on festive occasions to their children. These clothes are made up of expensive materials and excellent craftsmanship.

Different types of Chinese textile products

- Compound Fabric
- Bedding
- Chemical Fabric
- Feather & Down
- Fur Fabric
- Grey Cloth
- Adhesive bonded fabric
- Carpet
- Color Card & Color Comparator
- Functional Textile
- Glistening Fabric
- Industrial Textile

- Other Textiles
- Sack & Bag for Packing
- Silk
- Table Cloth
- Curtain
- Edging & Tassel
- Label & Tag
- Leather & Leatheroid
- Materials for Cap & shoes
- Textile Stocks
- Towel & Handkerchief
- Wool & Animal Hair
- Printed & Color-woven Cloth
- Sewing Thread & Tape
- Cotton
- Adhesive bonded fabric
- Decorative Cloth
- Embroidery & Drawn work
- Lace & Purfle
- Linen
- Military Material
- Textile Waste
- Twine, Rope & Netting

Carpets and other flooring traditions have been an essential part of Indian housing management from from ages. Indian households have used carpets and various kinds of floorings crafted in wool, silk, jute, coir, bamboo and wool. Till today. Indian carpets and floorings are enjoying patronage of large international clientele. Carpets and floorings are an integral part of Indian homes. Carpets are generally made from wool and silk. However, floorings like mats and durries are made up of a variety of material, ranging from wool, cotton, jute, coir, bamboo and grass. One can trace the origins of carpet weaving tradition in India to the Mughal dynasty.

Mughals not only used the Persian technique of carpet weaving, but were also influenced by traditional designs and motifs from Persia. Mughal carpets were as intricate as their miniatures and often depicted court life, animals and floral decorations. From the Mughals, the craft of weaving carpets spread to different regions of India, and the trend of using them became an international craze.

States like Kashmir, Uttar Pradesh, and others have their distinctive tradition of carpet weavings. Gabbe woolen carpets, chain stitch rugs and carpets, and namdas are also made in India. Durries are traditional Indian flooring and a cheap alternate to carpets. Durries come in a variety of designs and styles. Durrie weaving is common to many states. Mats are made out of a variety of natural material like, bamboo, grass, jute, coir etc. India has a varied mat making tradition. Bihar, Orissa, TamilNadu, Kerala, West Bengal and Northeastern states all have distinct mat weaving traditions of their own.

Indians have known weaving of material from Cotton since 5000 years. The traditional Indian cotton weaving revolves around 'Khadi'. Khadi is a cloth woven by hand using handspun yarn only. Fine cotton fabrics are also referred as Muslin. India has been famous for its ultra fine Muslins in the past. As handspun yarn is used in making Khadi, this activity is mainly carried out in the rural areas of the country.

Cotton weaving is the heart and soul of Indian textiles. There are 23 different varieties of cotton found in India and there are about 4 million handlooms producing cotton fabric. Cotton is used in producing a wide range of items like: saree, bed sheets, covers, napkins, shirts, summer wear, tablemats etc.

Cotton fabric is very popular in a tropical country like India, because of the soft twist imparted by the hand, maintains the hairiness of the yarn to an extent, which gives maximum comfort. Handloom producing Khadi weave cotton in such a way that the interlacing of threads provides maximum passage of air to the body, thus creating a cooling effect, making Khadi an ideal summer wear.

Every state has its own traditional cotton weaving tradition:

Fine embroidery and appliqué work done on garments, from various parts of Gujarat and mainly from Kutchh is done on cotton fabric.

Cotton is used to create beautiful saris and is an integral part of the saree weaving tradition of India. The tasteful varieties of sarees from Chander in Madhya Pradesh are not only produced in silk but also have silk- cotton types. The tribal regions of Madhya Pradesh are also known for their cotton sarees or Gamchhas.

The jamdani cotton sarees from Tanda and Banaras in Uttar Pradesh are popular. So are the tangail cottons Of West Bengal and Sambalpuri and Vichitrapuri sarees of Orissa. These saris come in a variety of colours and traditional motifs inspired by nature.

The city of Hyderabad is known for its ancient cotton weaving tradition. The Venkatgiri fine cotton sarees are produced here, which are ideal for summers. These sarees are mainly in off-white colour and decorated with dull golden motifs.

Andhra Pradesh is also known for Ponduru Khadi, which is obtained from wild variety of cotton. It is used in shirting and furnishings.

Cotton fabrics from the handlooms from West Bengal, Gujarat, Assam, Rajasthan, Karnatka, Madhya Pradesh, Maharashtra etc are used in creating a variety of items ranging from dress material, shirting, furnishing, hosiery, bedspreads etc.

The craft of weaving cotton fabric in India has achieved the status of small and medium scale industry. It not only caters to the local demands but is also exported to other countries on a large scale. Being eco friendly and 100% natural, the demand for products made out of cotton is on the rise.

African Textile

The weavers, dyers and other textile artists of Africa together makes an active contribution in creating exquisite and amazing range of textiles. African textiles usually embody a great variety of styles. Adinkara, kente and bogolan are some of the some of the African textiles which are becoming increasingly popular while some others like Yoruba, ase-oke and adire are equally beautiful but less well known.

In the ancient times the most important aspect of textiles or more precisely cloth in Africa was that cloth was used as form of money. The width of cloth strip was usually standardized in each region of Africa and therefore there used to be a regular number of such standard length cloth strips required to make a woman's wrapper cloth.

This would then be used to serve as the unit of value. Cloth was a convenient form of money primarily because it was used by everybody, fairly durable and easily sub dividable.

Though textile has great social relevance in Africa, but to the majority of textile workers, more than any other aspect of textile it is primarily an occupation.

African textiles usually use the following raw materials:

- Bast fibres
- Wool
- Cotton
- Silk
- Raffia
- Bark of certain trees

Various types of African textiles : An extensive range of hand made African textile follows :

- Adinkra
- Adire
- Akwete
- Aso-oke
- Bagolan
- Ewe
- Kente
- kuba
- Nupe clothes

African textiles are usually decorated either by dyeing the thread or the finished product.

African textile sector is busy getting influenced by the various developments in the global scenario and accordingly updating itself. Nowadays African designers are increasingly attempting to add a local appeal and flavour to their creations by using the tradition of weaving and textile design in contemporary dress styles.

Japanese Textile

Textiles offer a fascinating and revealing reflection of any society and this is particularly applicable in case of Japanese textiles. Starting with the modern age when Japan augmented its urbanization; textiles have served the purpose of acting as an emblem of social status and position. The style, colour motif and shape of the garment effectively depict the tradition and culture of Japan and at the same time also provides personalized information about an individual.

History of Japanese Textile In 1869 the capital of Japan was shifted from Kyoto to Tokyo and from this time onwards the

Nishijin weaving tradition seemed threatened with extinction. Though it had a negative impact on the weaving industry and consequently on the entire economy of Kyoto but the Nishijin weavers exhibited a remarkable spirit and flexibility in preserving their own craft and soon enough highly developed weaving technology and equipment was introduced.

The industry again started to grow along with Japan's new capitalist economy by 1890 when the Nishijin weavers embraced and applied modern technology to their own ancient and original textile art. The textile art of Japan particularly reached an epitome of excellence by exhibiting a cultural distinction and remarkable artistic skill in the Edo and succeeding Meiji periods (1868 - 1912).

Exclusive Japanese Textile Japanese textiles present an extraordinary and extensive collection of textiles. The rich heritage of Japanese textiles gets reflected through its traditional textiles like the kimono which is the traditional Japanese garment for men, women and children haori which is typical formal Japanese overcoat or obi which is a hand woven Japanese belt or sash tied around the waist or the fukusa which is an elaborately decorated and beautiful piece of cloth used for wrapping a gift while presenting it formally.

Exclusive Japanese Textile Art Tsujigahana textiles which belong to the period of fourteenth and early seventh century often reflect the epitome of excellence in Japanese Textile Art.

Tsujigahana textiles which were extremely creative and beautiful depicts the diverse changes in the history of Japan and these textiles were mainly utilized for making clothes, banners and such other items. The ornamental lavishness of the later Edo period often gets reflected through the best pieces of tsujigahana.

Southern Textile

Southern Textile is an organization that is involved in apparel/ general merchandise wholesale, liquidation and exporting business. Southern Textile was established in 1973 and it comprise of a team of experienced professionals who are operating all over the world and serve the buyers and customers of all the continents.

Southern Textile caters to the diversified needs of retail stores, wholesale distribution center or an e commerce market place as it offers a variety of branded products which are widely available for purchasing. The selected category comprise of computers, computer components, home office products, consumer electronics, digital cameras and video equipment, house wares, sports, hobbies, jewelry, gifts, accessories, apparel, shoes, furniture, domestics, home improvement, collectibles, and many more at reasonable low wholesale prices.

Southern Textile enjoys 30 years of experience and expertise in wholesaling and is devoted completely in finding the most excellent deals for its clients from all over the world.In this context it is necessary to clearly define and describe the term, Wholesale is a term that is closely associated with commercial aspect of textile product and textile industry.The practice of selling of textile goods or merchandise to retailers, commercial, institutional, or other professional business users or to other textile wholesalers is known as textile wholesaling. Wholesalers often physically assemble, sort and grade products in huge quantities, then break the bulk repack and redistribute them in smaller quantities. Southern Textile is one of the leading remarked of reverse supply chain merchandise to professional as well as wholesale buyers. This company enjoys 30 years of experience in liquidation industry. With over 30 years experience in the liquidation industry and it has long-term business relationships with major retailers, manufacturers and distributors. Southern Textile has expertise product range in major categories of consumer merchandise such as electronics, apparel, footwear, toys, furniture, entertainment and general merchandise.

The warehouse locations and national presence of Southern Textile offer professional purchasers fast and reliable access to lots of wholesale and discount goods to fulfill their requirements.

Southern Textile also has the credit of successfully using the Internet to generate an interactive market place which helps the professional buyers to access conveniently the brand name merchandise at concession in a short span of time.Apart from providing access to live inventory the website of this organization also offers the convenient and time saving option of placing online orders and arrangements for shipping requirements.

To take full advantage of online facilities, a business associate of need to have an account with Southern Textile and this account in turn offers the following access :

- Real time product availability.
- Real time order placement.
- Real time order status.
- Downloadable manifests.
- Quicker responses from sales representatives.
- Overview of recent online transactions.

Textile Converting

Textile Converting refers to the process of converting the textile fabrics to varying size and shape according to the requirement of the product. The product can be anything from a stylized garment to any home furnishing product. Textile Converting generally comprise of the following processes :

- Fabric designing based on the requirement of the product.
- Purchasing both dyed and undyed yarns and commissioning the knitting mills to knit the yarns into fabric according to the requirements of the specific product.
- Commissioning the finishers to finish the fabrics by using the process of washing, dyeing, bleaching and applying certain chemical finishes to the fabrics.

Chemical Conversion :

• Purified cellulose is chemically converted into a soluble compound in rayon production. A solution of this compound is passed through the spinneret to form soft filaments which are then converted or regenerated into almost pure cellulose. Rayon is termed as regenerated cellulose fibre due to the re-conversion of soluble compound to cellulose.

• Various types of rayon fibres are now a days used in commercial purposes and they are termed according to the process of conversion of cellulose into soluble form and consecutive regeneration. Rayon fibres are wet spun and it means that filaments developing from the spinneret pass directly into chemical bathing for solidification or regeneration.

Textile Printing

Textile printing is one of the procedure of textile decoration. **Textile printing** requires the correct coordination between the printing machinery support and the human skill for producing the best of the results. **Textile printing** uses various printing techniques of which screen printing is most predominant.

Screen printing was previously known as silk screening which refers to the printing technique that utilizes a stencil or a porous fabric to create a sharp edged image. Now a days it is very popular in both the fields of commercial printing and fine arts. Because of its advantage of low cost and ability to print on many mediums it is preferred over other processes like dye sublimation or inkjet printing. It started off during the First World War as an industrial process used primarily for printing flags and banners. Guy Mc Roy was the pioneer of screen printing technique. The greatest advantage of screen printing is that it possesses more versatility than any other traditional printing techniques.

- **Block Printing**–It was first developed in china. In this technique , first an original needs to be created ,then block with ink is used to create duplicates which involves a wide variety of skilled crafts men
- **Ink jet printing**–In this technique, the intended pattern along with the individual colour is formed by projecting small drops of ink which are special dye liquors in predetermined micro arrays onto the surface of the substrate.
- **Hand printing**–This is earliest form of printing used for the purpose of increasing the decorative value of textiles. This is the traditional technique of tie and dye.
- **Spray Printing**–In this technique, a spray gun is used to force the colour through a screen and electro-coating is used to apply a patterned pile.
- **Tie and dye**–This is particularly done on bright colours. This technique involves tying up of both wrap and weft threads where original colour needs to be retained and then dyed.

- **Kalamkari**–It is an exclusive and ancient craft done on textile fabrics. Kalamkari derives its name from the word Kalam which means Pen and Kari means Work and thus it means Pen Work. This art undergoes the laborious process of resist dyeing and hand printing.
- **Batik**–It is the procedure of dyeing the fabric by using the resist technique which covers the areas of the cloth to prevent it from colour absorption by using a dye resistant substance.
- Printing with the use of mordant is another technique of textile printing. Mordants are chemicals that absorb dye.

Phulkari Art

Sitting on the charpoys (beds woven with jute strings) pulled into the protective shade of a tree, or ensconced against a wall, women in villages and small towns all over Punjab are often busy creating spectacular **flower-embroidery** on dupattas, shawls or other garments. Called phulkari in local parlance, the origin of this beautiful art can be traced back to the 15th century AD.

The word phulkari literally means flowering. It is a form of craft in which embroidery is done in a simple and sparse design over shawls and dupattas. In some cases where the design is worked over very closely, covering the material entirely, it is called bagh (a garden of flowers).

The Making

The **embroidery of phulkari** and bagh is done in long and short darn stitch, which is created into innumerable designs and patterns. It is the skillful manipulation of this single stitch that lends an interesting and characteristic dimension to this needlework. While the stitch itself is uncomplicated, the quality of the phulkari depends upon the size of the stitch. The smaller the stitch, the finer the embroidery.

The threads used were of a silk yarn called pat. In the past, the silk threads were brought in from different parts of India, like Kashmir and Bengal and also from Afghanistan and China.

Using Colours : Bright colours were always preferred and among these, golden yellow, red, crimson, orange, green, blue, pink etc, were the popular ones.

For the embroidery, only a single strand was used at a time, each part worked in one colour. Shading and variation were not done by using various colours of thread. Instead, the effect was obtained by the dexterous use of horizontal, vertical or diagonal stitches. This resulted in giving an illusion of more than one shade when light fell on it and when it was viewed from different angles.

To keep the embroidered part clean while working on the cloth, the finished portion was rolled and covered with a muslin cloth. Specially created designs varied from village to village or region to region in Punjab and were given suitable names descriptive of their form. While phulkari was used to ornament cloth, the bagh ensured that not even a square inch of the base cloth was visible.

Base Cloth : The cloth primarily used and preferred by the women, was the home-spun, locally woven and dyed khadi. It was strong, long-lasting, and cheap and served the purpose of keeping the wearer warm during winters. Another reason was that the embroidery involved the counting of threads while doing the straight darn stitch. The coarse weave made this task easier. In addition, the thick cloth did not pucker and pull and could be worked upon without a frame. Usually, pieces of small width, about 45 to 60 centimeters, were worked on separately and the two or three strips were joined together to form the required size.

Repertoire of Motifs : Beginning with geometrical patterns, flowers and leaves, the repertoire of motifs was constantly enlarged. Birds, animals and human figures and objects of everyday use were inducted, along with vegetables, pots, buildings, rivers, the sun and the moon, scenes of village life, and other imagery. Phulkaris and baghs came to be embroidered in a stunning range of exquisite designs. In dhoop chaon, which literally means "sun and shade", an amazing interactive display of light and shade was created. The designs remained earthy and true to life. There was dhaniya bagh (coriander garden), motia bagh (jasmine garden), satranga bagh (garden of rainbow), leheria bagh (garden of waves) and many other depictions.

Today the most intricate and sought after phulkaris are the sainchi phulkaris, which bring scenes from rural Punjab to life. An incredible wealth of detail is embroidered onto cloth.

Reflecting Emotions : With time, the **phulkaris** became closely interwoven with the lives of the women of Punjab.

The joys, sorrows, hopes, dreams and yearnings of the young girls and women who **embroidered the phulkaris** were often transferred onto cloth. Many folk songs grew out of this expressive combination of skills and intense feelings. So, it is that one hears a young woman, whose betrothed has not sent a promised message to her, murmuring sadly, softly, as she *embroiders peacocks on a phulkari.* It was not long before phulkari folk songs became a part of the famous, pulsating folk dances of Punjab - the gidda and the bhangra.

A Symbol of Familial Ties : The women of Punjab created **phulkari** mostly for personal use. The cycle began with the young girl who followed her mother's chores and learned household work including this embroidery. When the girl got married, phulkari formed a part of her bridal trousseau. If a son was born to her, her mother would start preparing a vari da bagh, a gift she would present to her grand daughter-in- law.

The bagh was considered a symbol of marriage and among the wealthy families; sometimes up to fifty-one pieces of various designs were given to the bride. She, in turn, wore them for auspicious and ceremonial occasions. In some parts of Punjab, it was customary to drape the new mother with a bagh on the eleventh day after the birth of the child, when she left the maternity room for the first time.

Phulkari for Different Occasions : Phulkaris were also made for religious ceremonies or to be used at other festive times.

A phulkari is sanctified to be used as the canopy over the holy book of the Sikhs, the 'Guru Granth Sahib'.

For each different occasion, for each different setting, the versatile fingers and fertile imagination of the women of Punjab designed an impressive selection of phulkaris.

Different Varieties : There are different varieties of phulkaris and baghs made in Punjab.

The Chope, usually presented to the bride by her grandmother during a ceremony before the wedding, is embroidered with straight, two-sided line stitch and appears the same on the reverse. Only the border is embroidered and the center is left plain. It is usually red in colour and worn as a veil.

Vari-da-bagh (bagh of the trousseau) is also on a red cloth with golden yellow embroidery symbolizing happiness and fertility. The entire cloth is covered in a lozenge design with smaller ones within the border and is again intricately worked in different colours.

Ghunghat bagh or sari-pallu (covering for the head) has a small border on all four sides. In the center of each side, which covers the head, a large triangular motif is worked.

Bawan bagh (fifty-two in Punjabi) has as many geometrical patterns.

Darshan dwar (the gate offering a view of the deity) is usually for presentation in temples or to adorn the walls of the home when the Granth Sahib (holy book of the Sikhs) is brought to a house. The gate motif has been inspired by the arched verandahas of the temples. It is again, always on a red cloth. The architectural design depicts two outer panels of all gates with arched tops. The bases face each other with motifs of humans, animals, birds, flowers etc., giving the impression of passing through a crowded street.

Suber is a phulkari worn by a bride during marriage rites. It comprises five motifs, one in the center and one each in the four corners.

Chamba is a hybrid phulkari having a series of wavy creepers, stylized leaves and flowers. It came into vogue earlier this century.

Besides this, designs inspired by various day to day items, fauna and flora around us also found their way into this craft.

Surajmukhi (sunflower) is a cross between a chope and ordinary bagh in brilliant yellow, which creates a stunning effect.

Mor or tota is one that has a peacock or parrot motif.

Mirchi, as the name suggests, has chilies in red, orange or green usually on brown.

Belan and parantha symbolize the rolling pin and leavened bread.

Ikka or ace of diamonds has been inspired by playing cards.

Satrang is a seven-coloured phulkari.

Jewellery items like bangles, earrings, etc., are also embroidered At times, a snake was embroidered, guarding a woman's treasure. Another variety popular in Haryana was the Sheeshedar where small, round, matt-dull mirror pieces were included in the embroidered motifs.

Traditionally, the phulkaris and baghs were never sold in the markets as they were only woven by the women of the house for their personal use.

A New Form of Phulkari : A new form of phulkari is being embroidered these days.

It is not as detailed or time consuming as the older variety. Using a range of different fast coloured synthetic threads, it is embroidered from the top of the cloth rather than on its reverse.

The Punjab Government's Emporium, Phulkari, boasts of the best collection of this form of embroidery in the country. Their sizeable assemblage of a variety of these original pieces has been garnered from the villages in the state. Many of these are being exported, especially to the Middle-East.

5

Aspects of Textile

Textile Designer

A textile designer has to deal with a great range of duties and responsibilities. A **textile designer** is generally responsible for the following activities.

General work performed by textile designers :

- Attracting clients by proper planning and presentation
- Briefing the initial ideas and concepts to clients
- Analysis and research works
- Development of design
- Commercial aspects of finished product

Textile Designers primarily fall into two basic categories.

They either work for design agencies, manufacturers or retailers or they work as Freelancers.

Manufacturers

Textile manufacturer refers to organizations manufacturing textile products. Textile products include a wide variety.

Textile manufacturing organizations

Textile being a primary export item, lots of emphasis is given for growth and expansion in this sector. A textile manufacturing organization needs to concentrate on the following aspects.

- Adequate and necessary infrastructural support
- Specialized and efficient work team
- Smooth flow of raw materials
- High productivity
- Updated technology
- Good work environment

A textile designer needs the following skills and qualities:

- A basic understanding of colour, texture and pattern
- Creativity and artistic sense
- Adequate knowledge of various techniques and characteristics of different textile materials
- Knowledge of recent market trends
- Ability to work in a team
- Good communication skills
- Knowledge of relevant software used in textile designing
- Marketing, financial and administrative skills in case of freelancers

To create innovative and a wide variety of designs, textile designers usually works in knit, print or weave.

There are two basic fields in textile designing :

- Interiors (home furnishing)
- Fabrics for garments and accessories

Specialized works performed by textile designers are categorically listed below.

- Creating sample design set
- Liaising with clients to plan and develop designs
- Proper interpretation and representation of clients' ideas
- Creating sketches and designs for presentation to the clients
- Proper assessment of finished products
- Utilizing specialized computer aided design software (CAD) to create a great variety of designs

- Color, fabric and texture experimentation
- Possessing updated knowledge of new developments in the field of designing and production techniques
- Creating innovative design concepts
- Maintaining deadlines
- Attending trade shows regularly
- Getting updated with contemporary fashion trends in textile designing
- Developing and maintaining a network of business contacts

Textile Designs

Contemporary designs : Textile design requires special skills to create innovative designs.

The necessary skills follow:

- Creative ability
- Technical skills
- Commercial awareness of textile industry
- Research and data handling
- Critical evaluation and interpretation of materials
- Quick decision making ability

Key aspects of textile design : The core area of **textile designing** involves the following:

- Designing fabric by using different technique comprising of printing, weaving, ornamenting fabric, print technique, tracing embroidery and colour detailing.
- Providing support to the clients to visualize the design.
- Helping the clients to correct samples while executing prototypes.

Main activities in textile design : Textile designing involves the following basic activities.

- Creating sets of design samples
- Coordinating with clients to plan and develop designs
- Experimenting with colour, fabric and texture.
- Conceptualizing new innovative designs
- Attending trade shows regularly
- Keeping up with emerging fashion trends

Design studios : A textile design studio comprises of group of textile designers and is equipped with computers and design software to provide quality service and maintain professional standards.

Design products : Consists of bed linens, table linens, floor covering, cushions and curtains and many more.

- Textile designing
- Designing stalls for exhibitions
- Designing showrooms

A design studio ultimately puts up products for sale to prestigious buyers from all over the world.

Career in textile designing : With growing market potential for textile products there is a wide range of career options in this field. Textile design comprises both surface design and structural design.

Textile designers : Usually handle embroidery designs, print, weave and texture. Textile designers need to have detail information about textile fibres and different methods of textile design.

India has a rich history of textile design and a strong manufacturing base. The country's traditional, handloom sector co-exists with the highly mechanised mill sector and the decentralized power loom sector. The industry is also receiving global recognition with many international designers now sourcing fabric from India. Designers in the textile industry have a wide range of roles and responsibilities and are frequently required to make design decisions throughout the manufacturing process. Therefore there is a growing need for designers with technical and aesthetic proficiency. Design has emerged as a serious concern as industry strives to create a new value proposition for competitive niche positioning. Therefore there is a growing need for designers with technical and aesthetic proficiency.

LEVEL 1: Basic Knowledge, Skills and Concepts

Level 1 aims to develop sensitivity in design and acquire basic knowledge and skills for textile designing. This level will have modules in application of colour concepts & principles for: printing, weaving, knitting, & embroidery. The module will preliminarily emphasize on conceptualization of design ideas, processes & final execution.

LEVEL 2: Application and Practice

The program continues with professional approach for design development with a special focus on practical experience through industry related projects. This level will have design subjects with a blend of technology. Level 2 concentrates on design development of ideas, through an understanding of the commercial and technical aspects of the textile industry. This will equip the students to develop design competency skills as per the industry environment requirement. Year 3 provides an opportunity to students to study elsewhere in India or abroad at an approved textile institute. In this level a compulsory industrial training will familiarize the students with the actual atmosphere, pace and discipline of professional practice and the working culture of the textile industry.

LEVEL 3: Consolidation

Level 2 focuses on self-study under tutorial guidance, for an independent working culture. It aims to consolidate and implement the creative, technical and industrial experience. This level aims to produce a innovative, well-informed and motivated professional with the ability to analyze, synthesize and conclude various complex issues related to design, planning and technical management.

Thread Works : At the end of Level 3, students have the opportunity to exhibit their final collections in the public arena, where celebrated textile designers observe their creations; industry interests prospect creative talent for absorption in their design departments, or even engage them in freelance capacity. Top performers receive awards at the hands of well-known designers. All exhibitors benefit from the design insight shared by the accomplished visitors, a value that students are able to carry into their budding professional careers.

Careers : Graduates work as -

- Textile designers for apparel or home furnishing industry

- Designers for furnishing or floor coverings
- Fabric managers for textile industry
- Entrepreneurs
- Home stylists
- Design coordinators

Textile Fabric

Textile Fabric refers to the material used to prepare various categories and types of textile products for the textile industry. **Textile fabric** possesses an excellent national and international market. There are various types of textile fabric available for different products. Textile fabric can be broadly categorized as textile furnishing fabric and textile apparel fabric.

Types of textile fabric :

- Plastic fabric
- Chiffon fabric
- Cotton fabric
- Crepe fabric
- Furnishing fabric
- Industrial fabric
- Jacquard fabric
- Laminated fabric
- Narrow fabric
- Non woven fabric
- Quilted fabric
- Upholstery fabric
- Woven fabric
- Yarn dyed fabric

Uses of textile fabric : Textile fabrics are used for multiple purposes. The primary uses of textile fabrics follows.

- Used for making home furnishing products like bed sheets, cushion, cushion covers.
- Used for making carpets and rugs
- Used for preparing crests, emblems and ribbons
- Used for preparing embroidered apparel and garments
- Used for making jute products like bags, rugs and many more

- Used for preparing apparel and garments for men, women and children
- Used for preparing textile accessories
- Used for making embroidery threads and sewing treads
- Used for making towels like beach towels, bath towels, kitchen towels and aprons
- Used for preparing garments, blankets and woolen accessories
- Used for making yarns including cotton and knitting yarns
- Used for making mosquito nets, insect nets and bug nets
- Used for making tents and camps
- Used for making webbing belt, twine, rope and webbing straps
- Used for making plastic products

Textile fabrics are either natural fabric or synthetic. Our country is exporting natural textile fabrics since the ancient times. Synthetic fabrics are comparatively new and evolved with the continuous growth in textile industry. According to the need and constant innovation in textile fabrics, water proof, oil resistant, UV resistant as well as abrasive resistant fabrics are also available now a days.

Textile Art

Like other art forms textile arts involves creativity and innovative ideas and the necessary skill to implement it. **Textile arts** simply crafts the uniqueness in a given fibre product either its natural fibre or synthetic fibre.

Every corner of our country contributes in making a colourful history of textile arts. Though a contemporary textile art has undergone a vast change in various aspects but still its origin can distinctly be traced back to ancient times. Textile is in essence a medium for acquiring the profession that serves the people.

Textile arts comprise of incorporating beautiful images and colour elements along with an integrity of exquisite designs

Textile arts include a large number of forms and materials.

- Jute handlooms
- Silk fabric
- Khaddar fabric

- Tie and dye materials

Jute Handloom : Jute is called the "Golden Fibre" and it is the second most important fibre after cotton. It's a Lignocelluloses fibre which means its partially a textile fibre and partially a wood.Botannically *it falls under bast fibre* category. Jute and coir crafts are relatively new compared to other traditional craft traditions of India. These fibres are mainly used in the production of utilitarian items, though some decorative items are also made.

India and Bangladesh are the main producers of jute and coir. Indian jute handicrafts are popular over the world. Skilled artificiers craft numerous items of daily use from jute fibre.

A Jute item is gaining popularity because it's the cheapest known natural fibre and is 100 per cent eco-friendly. Since early nineteenth century Bengal has been producing maximum quantity of Jute all over the world. West Bengal is the home of jute handicrafts from India. Craftsmen here create a myriad of utilitarian items made from jute. An elegant range of home décor and furnishings like cushion covers, table covers; tea cozies etc are available in jute fibre. Exquisite and colourful jute floorings are also made from jute. Garments made out of woven jute fibre are gaining popularity. Jute fabrics are decorated with tie and dye, embroidery and block printing. Jute is also used for making baskets, mats and beautiful toys.

The discovery of polythene and nylon in 1970s inflictec a huge loss to the jute planters. But since the notion of ecofriendlines has emerged jute is again in high demand and farmers have begun to earn some money. Madhya Pradesh is also known for Jute items from Bhopal, Indore and Gwalior, which include hanging lamps, baskets, flower vases, purses, table mats and footwear etc.

Jute has got immense importance for packaging purpose. Cereals, vegetables, cement and numerous other things are packaged in the sacks made by jute.

Similarly Coir is also an important material which is not only eco-friendly but also cheap and very useful for innumerable purposes.Coir is obtained from coconut husk. Kerala the 'land of coconuts' is the largest producer of coir and accounts for about 75 per cent of the total coir production from India. It is also known as the wonder fibre.

A variety of products are made from coir. Floorings, mats, carpets, bags, wall hangings etc are made from coir fibre. Coir foam or rubberized coir is used in making mattresses and cushions. Products made from coir foam have gained popularity because of their softness and durability. The unique feature of this product is that it stays cool in summers and warm in winters.

Silk Fabric :

History of Silk : Since the 27 th century B.C., the Chinese have produced and used silk fabric. In fact, raising silk worms was one of the many chores of the farm women in China. From China, silk was exported via the Silk Route.

The Chinese never let out the secret of how the silk was produced. However, in later years, Christian monks smuggled the eggs out of the country; hence introducing silk manufacture in other nations as well.

The Making of Silk Fabric: The process of making silk is delicate and involves a number of steps :

- The first stage is called sericulture. This is the cultivation of the silk worms. The most popular species for obtaining mulberry silk is Bombyx mori. The worms are raised in a controlled environment and are fed mulberry leaves. The worms form a cocoon around themselves by secreting a protein from the top of their head.

  The farmers collect these cocoons and deliver them to the factory, where they are subject to filature operations.
- The first step is to sort the cocoons according to colour, size, shape and texture.
- Then, the cocoons are made to go through a serious of hot and cold immersions. In this way, the sericin (the gummy substance that holds the fibroin strands in the silk filament together) is softened.
- Once this is done, the filament is unwound from the cocoon and combined to produce a thread of raw silk. This is the process of reeling. Usually, three to ten strands are reeled at a time.

- Finally the skins, into which the filament was reeled, are packed into bundles called books which are then put into bales to be exported to the mill.
- In the mill, the silk fibre is woven into silk fabric, using either a hand loom or a power loom.

Indian Silk Fabric:

India is the second largest producer of silk after China and the largest consumer of silk in the world. As per the 2001-02 records, India produced 17550 MT of silk.

Types of Silk Fabric:

There are innumerable varieties of silk moths, but only four main types of commercially valuable natural silk:

- ***Tasar silk***: This is a copperish coloured silk. It is coarse and is used mainly for furnishings. It is produced by the silkworm *Antheraea mylitta,* which mainly thrives on the plants. Asan and Arjun. It is reared on trees in the open.
- ***Muga silk***: This is a golden yellow coloured silk that is produced in Assam. It is obtained from the semi-domesticated silkworm, *Antheraea assamensis* which feeds on the aromatic leaves of Som and Soalu plants.
- ***Eri silk***: This is got from the domesticated silkworm, *Philosamia ricini* that feeds mainly on castor leaves. The eri cocoons are open-mouthed.

In India, mulberry silk is produced in many states such as Karnataka, Andhra Pradesh, Tamil Nadu, Jammu & Kashmir and West Bengal. The other types of silk are produced in Jharkhand, Chattisgarh, Orissa and the north-eastern states.

There are a wide variety of silk textiles made from these four kinds of natural silk. These may be handwoven or woven in power looms. In the mulberry silk variety, some of these silk fabrics are:

- ***Plain silk***: This variety of silk textile may be produced on both the hand loom and the powerloom. Available in various shades and patterns, and an entire range of qualities.
- ***Organza***: This is a very thin silk cloth that is made of highly twisted yarn.
- ***Crepe***: This silk textile is woven from 's' and 'z' twisted yarn. The crepe sarees of Mysore are very beautiful.

- ***Satin***: An extremely elegant type of mulberry silk fabric. Banarasi satin sarees are very popular.
- ***Matka silk***: A thick kind of silk fabric, used for furnishing. By varying the amount of yarn used, the texture and thickness may be modified. It is produced in Bhagalpur in India.
- ***Murshidabad silk***: Produced in the Murshidabad district of West Bengal, this type of silk cloth is used to make sarees and scarves. It is available in varying qualities.
- ***Dupion***: This type of Indian silk fabric is a speciality of the handlooms in Bangalore, India. It is available in a number of shades; and used for both garments as well as soft furnishings.
- ***Charka silk***: This is a thick variety of silk from India that is used to make *zari* sarees. It is woven on a handloom.
- ***Chiffon***: A very soft and light variety of silk fabric, it is made of highly twisted yarn that is woven on a power loom. It is a very thin, but strong fabric.
- ***Chinnon***: This is a very soft and crimped silk textile. It is woven of twisted yarn on a power loom.
- ***Tabby silk***: This variety is produced in Kashmir and is used to make sarees and scarves.

Textile arts include various activities and art forms.

- Crochet
- Embroidery
- Knitting
- Lace work
- Needle work
- Weaving

Women in **textile arts** form a pivotal role in not only providing a creative vent out but it also makes them financially independent. With the increasing scope in this field, many women are consciously opting out for textile arts as a career choice.

Products in textile arts are often exquisite art forms which deserve to be preserved. But it also serves as wonderful gift items for our near and dear ones. Such gift item usually includes wall hanging, bags, accessories, fashionable utensils, stylish fabric.

Textile Machinery

History of textile machinery : Textile Industry its evolution and progress forms an integral part of the history of textile machinery. Since the dawn of civilization, clothing was one of the man's primary needs. This led to the spinning of fibre into yarn and the cloth weaving which finally resulted in innovation of new technologies for textile industries.

Early spinning The first textile machinery used was the spinning wheel. It first developed in India and then in 14 th century it reached Europe.

History of loom : Loom is ancient in origin and the modern invention to increase its skill was the flying shuttle which John Kay patented in 1733.

Textual Mechanization : The initial enhancement in the early spinning machines took place in 1737 when Lewis Paul and John Wyatt discovered the roller method of spinning jenny and water frame by Samuel Crompton in 1779.

Types of Textile Machineries Textile Machineries comprise primarily of two types:

- Textile process machineries
- Textile working machineries and equipments and accessories

Various Textile process machineries :

- Cloth finishing machines
- Knitting machines
- Fabric seaming machineries
- Crochet machines
- Lace making machines
- Label making machines
- Quilting machines
- Textile finishing machines
- Textile sourcing machines
- Textile spinning machines
- Textile winding machines
- Textile edge control device
- Thread winding machines

- Tufting machines
- Weaving machines
- Zipper making machines
- Woolen mill machines

Various Textile working machineries and equipments and accessories :

- Applique scaling machines
- Attaching machines
- Cloth measuring machines
- Cloth cutting machines
- Embroidery machinery
- Garment machinery
- Industrial sewing machine
- Laundry dryers
- Monogramming machines
- Textile bleaching machines
- Textile folding machine
- Textile trimmers machine

Uses of Textile Machineries It is primarily used in cotton mill, covering plants, wool mill, garment factory; man made factories of fur and trades goods inspection units for the entire length of fabric rolling. It is especially beneficial for inspecting and rolling export fabrics.

Textile Consultants

Textile Consultant offers technical, engineering creative, marketing and other related services in the textile industry.

In simpler words, Textile consultants offer solutions to business needs in textile industry. Textile consultancy consists of a number of services which are extremely beneficial for the manufactures in the textile sector Labour issues.

Primary Services provided by Textile Consultants Product or process development

- New process and product research
- Detail analysis of technical specifications
- High level of technical assistance

- Infrastructural changes

Troubleshooting and Process Audit :

- Fibre and fibre processing audit, scouring, spinning weaving, dyeing and finishing procedures for improving the quality of the product and making it cost effective.
- Identifying and immediate rectification of the problems related with textile processing.

Designing and assessment of equipment Processing sequences :

- Defining properly and adequately the dyeing and finishing requirements of the specific quality of fabric.
- Factors that the apparent shade of the fabrics.
- Choosing the right kind of dye to meet specific requirements.
- Creating new innovative fabrics through chemical and mechanical finishing.
- Equipment and capacity issues.

Studying and properly assessing the current work procedures :

- Work practice comprise of instructions for current work, dye house procedures.
- Variables in the process of dyeing.
- Rationalization of dyes and chemicals.
- Labor issues.
- Issues involving technical expertise.

Textile Wholesaler and Retailer

Textile Whole sale and Retail are terms closely associated with the commercial aspect of textile industry. The practice of selling of textile goods or merchandise to retailers, commercial, institutional, or other professional business users or to other textile wholesalers is known as textile wholesaling. Wholesaler often physically assemble, sort and grade products in huge quantities, then break the bulk and repack and redistribute them in smaller quantities

The sale of textile goods or merchandise usually for personal or household usage either from a fixed location or from associating subordinated services is known as textile retailing. Textile retailers usually refer to those groups of who purchases huge quantities of textile products from manufacturers or importers either directly or through a wholesaler and ultimately sells individual product item or small quantities to the general customers in shops or stores.

Textile products

- Apparels for men, women and kids
- Backpacks
- Bags and cases
- Lace and trimmings
- Leather products
- Lingerie and underwear
- Linen wear
- Carpets and rugs
- Chemical fabrics
- Cashmere and mohair wear
- jackets
- Purse and wallet
- Table Cloth
- Cotton fabrics
- Grey fabrics
- Table cloths and linens
- Textile accessories
- Printed fabric
- Yarn
- Gloves and mittens
- Textile house hold products

The practice of selling of textile goods or merchandise to retailers (Textile wholesale and retail), commercial, institutional, or other professional business users or to other textile wholesalers is known as textile wholesaling. Wholesalers often physically assemble, sort and grade products in huge quantities, then break the bulk and repack and redistribute them in smaller quantities.

Textile Wholesale Products–The textile sector is large and comprise of a wide variety of textile products.

Some of the primary textile products associated with textile wholesaling is categorized below:

- Apparels for men, women and kids
- Backpacks
- Bags and cases
- Lace and trimmings
- Leather products
- Linen wear
- Carpets and rugs
- Chemical fabrics
- Cashmere and mohair wear
- Purse and wallet
- Reflective materials
- Home furnishing products like cushions, cushion covers, pillows, table cloth, mats etc

Textile Artist

A textile artist can be defined as an artist who is engaged in creating his or her art by utilizing different types of fibres, threads or fabrics. Textile artists forms an intrinsic part of textile tradition of any country. Textile Tradition is neither static nor geographically rigid and along with constant interaction between diverse cultures, the textile tradition gets influenced by each other. The textile art of various nations originates from indigenous textile crafts and reflects local cultural tradition and depicts its social significance.

Most of the countries possess a particular textile trait and textile artists of that country primarily concentrate on this trait by either enhancing its creativity or functional aspect or creating innovative textile fabrics and products by combining these traits. Textile artist not only incorporates beautiful images and colour elements but also creates integrity of exquisite designs.

Textile Fabric can be broadly categorized as textile furnishing fabric and textile apparel fabric. Textile Artists usually deals with a single or combination of these following fabrics:

Types of textile fabric

- Cotton canvas
- Plastic fabric
- Chiffon fabric
- Cotton fabric
- Crepe fabric
- Furnishing fabric
- Industrial fabric
- Jacquard fabric
- Laminated fabric
- Narrow fabric
- Non woven fabric
- Quilted fabric
- Upholstery fabric
- Woven fabric
- Yarn dyed fabric

A textile artist needs the following skills and qualities :

- A basic understanding of colour, texture and pattern
- Creativity and artistic sense
- Adequate knowledge of various techniques and characteristics of different textile materials
- Knowledge of recent market trends
- Ability to work in a team
- Knowledge of relevant soft wares used in textile designing

Textile artists are usually engaged in textile apparels or textile furnishings. Distinct style and tradition created by Indian textile artists

- Silk and cotton weaving predominates the rich tradition of weaving in India. Silk weaving is most popular in various parts of the country. Assam, Banaras, Mysore, Surat, Kanchipuram and Surat are all important centers of silk weaving. All these centers specialize mostly in cotton and silk saree weaving and some of the popular traditional Indian saris are Banarasi, Patola, Baluchari, Pochampalli, Paithani and many others.
- The ornate style of appliqué most commonly done bright coloured fabrics from the Kutchh region of Gujarat is very popular in the country as well as in abroad.

- The state of Punjab is famous for its phulkari work.
- Lucknow, a city in Uttar Pradesh is famous for its chikan work.
- The hilly region of the country produces a rich variety of woolen textiles. The pashmina and shahtoosh shawls of Kashmir, shawls and woolen garments of Himachal Pradesh and other north eastern states provides excellent examples of world famous woolen Indian textiles.
- The home furnishing utilitarian textile products like bed spreads and sheets, pillows and cushions, linens and mats, curtains and napkins, carpets and rugs and many such other items are created by various textile artists from different parts of the country.

Textile Weaving

History of Textile Weaving Weaving is acknowledged as one of the oldest surviving craft in the world. The tradition of weaving traces back to Neolithic times that is approximately 12,000 years ago. Even before the actual process of weaving was discovered, the basic principle of weaving was applied to interlace branches and twigs to create fences, shelters and baskets for protection.

Weaving is one of the primary methods of textile production and it involves interlinking a set of vertical threads with a set of horizontal threads. The set of vertical threads are known as wrap and the set of horizontal threads are known as weft. Weaving can be done by hand or it can also be done by using machines. Machines used for weaving is termed as loom.

Loom originated from crude wooden frame and gradually transformed into the modern sophisticated electronic weaving machine. Now a days weaving has become a mechanized process though hand weaving is still in practice.

Initially weaving was done by hand by the weavers but with the mechanization of the weaving and industrial revolution, the traditional hand weavers felt threatened: a hei very livelihood

Wool available from sheep served as the earliest textile fabric available for spinning into yarn and then for weaving into cloth.

The first spinning wheel was discovered in India and then it reached Europe in the latter half of the 14th century.

Modern Machines Used in Weaving. The impact and wide spread use of modern, technically updated machinery becomes clearly evident list of following list of weaving machineries :

- Rapier loom
- Collar making machine
- Lace braiding machine
- Air jet loom
- Jacquard weaving machine
- Computerized jacquard ribbon loom
- Glove weaving machine

Textile Mills

Textiles are formed after the process of weaving, knitting, crocheting, knotting and pressing fibres together. Textile Mills refers to manufacturing plants for making textile fabric and products.

Textile Mill Industry is one of the largest industries in India.

It approximately provides employment to almost 10 lakh workers. A huge number of associated industries also depend on the textile sector. The Indian textile mill industries went through a process of phenomenal growth for the past four decades.

Conditions of the Indian Textile Mills : For the past few years the sickness and consequent closure of textile mills has been a matter of great concern in our country.

The primary reasons behind this are:

- The primary reason behind the sickness is structural transformation leading to composite units in organized loosing ground to power looms in decentralized sector due to greater cost effectiveness.
- Low productivity due to lack of adequate modernization
- Stagnation in demand
- Inability of certain units to expand in national and international market
- Increase in cost of input
- Inadequate working capital

Government of India took initiative to take adequate measures to tackle this problem.

- Board for industrial and Financial Reconstruction (BIFR) was set up to detect sickness and the sick companies.
- Textile Workers' Rehabilitation Fund Scheme (TWRFS) was set up to tackle the problems.

Few Leading Indian Textile Mills :

- Adarsh Textile Mills: Manufacturer and exporter of good quality woolen and synthetic blankets
- Amritsar Swadeshi Woolen Mills: Pioneer in manufacturing heavy woolen yarn and largest manufacturer of fabric
- Aroon Mills: Manufacture of textile auxiliaries
- Mohan Thread Mills: Manufacturer of high quality embroidery yarn and threads

Products of Textile Mills : Textile Mills usually produces both yarn products and fabric products.

Yarn Products

- Cotton yarns
- Blended yarn
- Synthetic yarn
- Specialty yarn

Fabric Products

- Woven fabrics
- Knitted fabrics
- Grey and dyed fabrics

Textile Dictionary

The word 'dictionary' refers to a book containing an alphabetical list of words, with information given for each word, usually including meaning, pronunciation, and etymology or it may be a book which is comprised of the words of a language with translations into another language or it can also be a book listing words or other linguistic items in a particular category or subject with specialized information about them.

Originally dictionaries were mainly available in printed versions like books but with the technological development the online versions of dictionaries have also become very popular to the users. The online dictionaries are not only cost effective but they are also more easily accessible and user friendly. Textile Dictionary refers to the dictionary which deals with textile and textile related words and terminologies as well as information on various aspects of textile.

Elsevier's Textile Dictionary is available in English, Italian, French and Spanish and this dictionary covers information on textile machinery and equipment; textile fibres and goods; textile operations and processes; textile chemicals and tests; sewing, knitting and embroidery; making-up, fashion and accessories; colours and patterns; business and trade.

Utility of Textile Dictionary

- Access to vast range of information
- Easy up gradation of personal knowledge
- enhancing professional accuracy
- Exchange of opinion and information through some of the online dictionary sites
- Clarification of doubts and confusions leading to better productivity.

6

Fields of Textile

India has a diverse and rich textile tradition. The origin of Indian textiles can be traced to the Indus valley civilization. The people of this civilization used homespun cotton for weaving their garments. Excavations at Harappa and Mohen -Jo-Daro, have unearthed household items like needles made of bone and spindles made of wood, amply suggesting that homespun cotton was used to make garments. Fragments of woven cotton have also been found from these sites.

The first literary information about textiles in India can be found in the Rig-Veda, which refers to weaving. The ancient Indian epics-Ramayana and Mahabharata also speak of a variety of fabrics of those times. The Ramayana refers to the rich styles worn by the aristocracy on one hand and the simple clothes worn by the commoners and ascetics.

Ample evidence on the ancient textiles of India can also be obtained from the various sculptures belonging to Mauryan and Gupta age as well as from ancient Buddhist scripts and murals (Ajanta caves). Legend has it that when Amrapali, a courtesan from the kingdom of Vaishali met Gautam Buddha, she wore a richly woven semi transparent saree, which speaks volumes of the technical achievement of the ancient Indian weaver.

India had numerous trade links with the outside world and Indian textiles were popular in the ancient world. Indian silk was popular in Rome in the early centuries of the Christian era. Hoards of fragments of cotton material originating from Handicrafts have been found in the Egyptian tombs at Fostat, belonging to 5th century A.D. **Cotton textiles** were also exported to China during the heydays of the silk route.

Silk fabrics from south India were exported to Indonesia during the 13th century. India also exported printed cotton fabrics or chintz, to European countries and the Far East before the coming of the Europeans to India. The British East India Company also traded in Indian cotton and silk fabrics, which included the famous Dacca musl. Muslins from Bengal, Bihar and Orissa were also popular abroad.(Muslin-a very thin cotton material) (Chintz-cotton cloth, usually printed with flowery patterns, that has a slightly shiny appearance)

The past traditions of the textile and handlooms can still be seen amongst the motifs, patterns, designs, and the old techniques of weaving, still employed by the weavers.

Home Textile

The term 'Textile' is a Latin word originating from the word 'texere' which means 'to weave'. Textile originally referred to a woven fabric however, gradually the term textile as well as the plural textiles implies fibres, filaments and yarns. Home textiles refer to the textiles used for home furnishing. It comprise of extensive range of functional as well as decorative items or products used mainly for the purpose of decorating our homes.

Textile home furnishing fabrics or home textiles consist of both natural and man-made fabrics. At times, these textile fabrics are made strong and durable by blending them.

The **textile fabrics** most commonly used in home furnishing are listed below.

- Silk fabric

- Cotton fabric
- Jute fabric
- Rayon fabric
- Wool fabric
- Nylon fabric
- Polyester fabric
- Plastic fabric
- Leather fabric
- Satin fabric
- Organza fabric
- Organdy fabric

Different techniques are used by home textile manufacturers to enhance the decorative aspect of the home furnishing products.

The basic and most common techniques are categorized below.

- Printed
- Embroidered
- Woven
- Non woven
- Dyed
- Stitched
- Hand made

Home textiles include a wide range of products. Some of the most common home textile products are listed below.

- Home furnishing fabrics
- Bed spreads
- Blankets
- Pillows and pillow covers
- Cushion and cushion covers
- Carpets and rugs
- Wall hangings
- Different types of towels
- Table cloth and mats
- Kitchen linen and other kitchen accessories
- Bathroom accessories

Bedspreads

It is commonly held that the home is an expression of oneself. Perhaps because this is where a person spends most of his or her time. No wonder then that homemakers the world over take great care in setting up their homes, furnishing and decorating it. And this is no easy feat, for the variety of home décor items available is immense. Take for instance, the Indian bedspread.

Indian bedspreads are celebrated for their beautiful colours, fine embroidery and prints. They are available in almost every fabric, including cotton, satin, silk, polyester, woolen and various blends. And they are adorned with magnificent patterns that would make any homemaker proud.

Styles of Decorating Bed Spreads : Block Printing is an art that is practiced in several parts of India. Artisans use wooden or metal blocks that have been engraved with lovely designs, to print colourful patterns. Bedspreads from India are very often block printed, with motifs of flowers, leaves, figures and geometric patterns.

Painted Indian bedspreads are also a visual treat. Batik, tie-dye and kalamkari are just some of the painting styles that artisans use to beautify bed covers.

The technique of batik involves the resist-dyeing method, wherein wax is used to cover the portions that are not required to be coloured, and the cloth is then dipped in dye.

In the tie-dye style, portions of the cloth are tied with thread in interesting ways, and then the cloth is dyed. When the threads are removed, beautiful patterns are formed of the tied areas.

The art of Kalamkari is a combination of hand block printing and painting using a brush-like pen. The two major centers of Kalamkari in India are Sri Kalahasti and Masulipatnam.

The Masulipatnam designs are Iranian in character; the most popular motifs are Persian motifs like trees, creepers, flowers and leaf designs. In Masulipatnam, Kalamkari work is mainly done on bed covers, curtains and garments, using a combination of wooden block printing and hand painting.

In Sri Kalahasti Kalamkari work, temples are a major source of inspiration. It was because of the demand for scrolls and wall hangings with Hindu mythological stories that Kalamkari flourished in this village. These themes are painted in the panels, and there is a script painted along the border. The popular motifs are Hindu gods and goddesses. The work is done entirely with a brush-like pen.

Nothing can beat the beauty of a skillfully embroidered bedspread. And when it comes to embroidered Indian bedspreads, one is easily spoiled for choice.

Kantha Embroidered Bedcovers: The beautiful Kantha embroidery of West Bengal uses the quilting or running stitch to create designs of gods and goddesses, flowers, animals and geometric patterns.

Kashmiri Embroidered Bed spreads :

- The Kashida embroidery of Kashmir gets its inspirations from nature. Employing no more than two different stitching styles, pretty patterns of flowers, creepers, leaves and mangoes are created.
- The hook or ari embroidery of Kashmir is used to create floral patterns in chain stitch.
- Kashmir's papier mache embroidery is done in bright colours using satin stitch. The patterns are usually floral and outlined in black.

Phulkari Embroidery on Bed covers: The Phulkari style of Indian embroidery hails from Punjab. Executed on handspun khadi cloth, it uses simple darning stitches. Floral patterns in bright colours are the trademark motifs.

Banjara Embroidery on Indian Bedspreads: This embroidery is unique to the Banjara nomadic tribes of India. It is embellished with mirror and bead work, shells, ivory and colourful threads.

Banni Embroidered Bedspreads from India: Banni embroidery is done with silk floss. It is brightly coloured and uses mirrors and beads. It hails from Kutchh in Gujarat, India.

Mirror Work Embroidery on Bedcovers: In sheeshedar or mirror-work embroidery of Gujarat and Rajasthan, glass is cut into different shapes and embroidered in the fabric. This embroidery is characterized by a combination of cross, satin and button-hole stitch, along with the mirrors. Apart from painted, printed, and embroidered bedspreads, India bedcovers that are quilted and woven are also well known.

Indian Carpet Weaving Tradition

Weaving carpets has been a vocation in India since medieval period. The Indian carpet traces its provenance to the reign of the Mughals. Legend says that when Babur came to India, he missed the luxuries of Persia. In later years, in around 1580 AD, Akbar invited some Persian carpet weavers to his court. This was the beginning of Carpet weaving in India. The local artisans picked up the technique and designs from the Persians, and carpet weaving centers were set up in Agra, Delhi and Lahore.

Ever since the first carpet was woven in India, the carpet weaving industry has grown by leaps and bounds. Today, carpets from India are woven by hand or in a power loom. Though silk carpets and Indian wool carpets are the most popular, cotton, jute, bamboo, grass and coir are also used to make these attractive Indian floor coverings.

Indian Carpet Designs

The motifs on Indian carpets are an exquisite blend of original Persian patterns, and ethnic Indian designs. Some of the popular patterns woven into carpets are:

Kashan: This is an interconnected leaf and tendril design. There is no central medallion.

Tabriz: This is a leaf and tendril pattern, with an intricate motif at the corners and a floral border. There is an elongated medallion.

Ispahan: This is a leaf and tendril pattern with intricate corners, a floral border, and a round medallion.

Bokhara: This design consists of repeated oval medallions that are called gulls or 'elephant footprints'.

Herati: This pattern comprises diamond shapes and leaves. There is usually no medallion.

Heriz: This design used in Indian carpets is geometric. There is a medallion resembling a starburst too.

Serapi: This pattern evolved into the heriz design described above. It is geometric and consists of a medallion from which arms extend.

Mir: Also referred to as Boteh, this Indian carpet design is a paisley pattern spread all over.

Kerman: This consists of a floral medallion, intricate corners and a border.

Serouk: This is simply a pattern of independent flower bouquets.

Mahal: This is an abstract floral design, markedly geometric.

Tribal: These are irregular patterns of various kinds.

Aubusson: A large floral medallion, intricate corners, an elaborate border. This is a French design.

Savonnerie: This is an elaborate Aubusson pattern.

Bidjar: There is a hexagonal medallion, and a herati field surrounding it.

Indian Silk Carpets : These traditional Indian carpets are procured from Kashmir. The silk carpet of Kashmir is hand knotted, and is renowned for its softness and the superb skill with which it is crafted.

The type of silk carpet created depends on the yarn used. Silk carpets of India are created on a silk pile, a cotton pile or a woolen pile. The 100 % pure silk carpets have between 400-1600 knots per square inch. This high number ensures their durability.

The most popular motif in Kashmiri carpets is the tree of life. Other common subjects are hunting scenes, bird and animal motifs and historic subjects. The carpets of Kashmir are available in a number of colours. Red, blue and yellow mineral hues are the most common.

Industrial Textile

Textiles which are specifically designed and engineered to be capable of using in products, processes or services of mostly non-textile industries are termed as industrial textile. Another distinguishing feature of industrial textile is that unlike the traditional use of ordinary textiles in apparel and furnishing, it is primarily utilized by industrial professionals in different types of high performance and heavy-duty applications from non-textile industries. Traditional textiles enjoy a distinct form of textile art and textile design.

The term industrial textile (textile industry) is mainly used to refer non-traditional textiles and it is also known as technical textiles or high performance textiles or high tech textiles. According to the definition industrial textiles possesses three primary uses.

- It can be utilized as a component part of any other product and provide direct contribution in strength, performance and other related properties of that product. e.g. tire cord fabric in tires.
- It can be used as a device in the procedure of manufacturing another product. e.g. paper machine clothing in manufacturing of paper.
- It can alone be utilized to perform a single or multiple specialized functions.

History of Industrial Textile

Traditional textiles are considered to be a little older than industrial textiles though both of their origin trace back to several thousands years ago. The tradition of modern industrial textile started with usage of canvas cloth to sail ships in the ocean. The industrial textile market witnessed a remarkable change in the first half of 20th century with the invention of man-made fibres.

Usage and Significance of Industrial Textile :

- It plays a significant role in space exploration and space suits are created with layered fabric system to provide protection to the astronaut.
- Space vehicles are made up of specialized industrial textiles as it can withstand a temperature of several thousands degree.
- Industrial textiles are widely used in military applications to provide greater protection to the soldiers
- Specialized industrial textiles are used in aero planes and environmental protection and high way construction.

- Industrial textiles even makes contribution in making artificial kidneys used during dialysis of patients. The greatest advantage of industrial textile is that it not only possesses strength, flexibility and elasticity but also offers various designing options.

Standard Textile

Standard Textile was established in 1940 and presently it not only sells products domestically but also to customers of 32 countries from different parts of the world. **Standard textile** is an organization which excels primarily in service, quality, innovation and cost effectiveness. Standard textile is acknowledged as one of the global leaders in manufacturing healthcare, hospitality and institutional textiles, apparel, surgical, incontinence care, decorative products and linen management services. It is a vertically integrated company that is involved in managing its own extensive research and development activities, manufacturing and worldwide distribution.

The product range of Standard Textile is wide and diversified.

Decorative Products : All the innovative decorative products of standard textile of Standard Textile are aimed and designed to meet the specific need for healthcare, hospitality, educational and commercial.

- Cubicle Curtains / Cubicle Curtain Tracks & Hardware
- Bedspreads
- Solar Shades / Mini & Vertical Blinds
- Replacement Cushions
- Top Treatment
- Drapery

Health care products and services

Healthcare Services includes the following

- Clinical Services
- Commercial Laundry
- Consultative
- Control Tex-Linen Management Programs

- Customer Service
- Industrial Engineering
- PrePak Surgical Linen Management System
- Technical Services

Healthcare Products includes the following

- Apparel
- Sheets & Pillowcases
- Terry Products
- Surgical Services
- Blankets & Bedspreads
- Hamper bags
- Support Service Products
- Mattresses Pads
- Pillows
- Bibs

Hospitality Products and Services

Hospitality Products includes the following

Terry Products

- Centium Core Technology sheet
- Chamber loft Pillow
- Custom Jacquard Terry
- Euroclassique
- Eurosoft
- Euro spa pool towels
- Euro touch
- Impact shower curtains
- Luxury stripe
- Room ready for you
- Wash towel
- Room ready value

Sheeting

- Centium satin
- Comfort twill
- Easy id info sheet
- Truval sheeting

- Truval laundry comparison
- Upgraded single pick percale

Blankets and Top of the Bed

- Acrymax blanket
- Bed scarf and bolster
- Essential elements
- Herringbone performance blanket

Hospitality service of Standard textile comprise of Textile-Online Ordering System

Work Wear Products : Standard Textile provides a great quality and range of work wear products. The work wear products comprise of the following.

- High performance coverall
- Lined hip jacket
- Performance lined sport jacket
- Unlined sport jacket
- Utility liner

Global Presence of Standard Textile

Manufacturing and Technical capabilities of Standard Textile is most advanced in the world and it includes various aspects from raw materials sourcing and manufacturing to a global logistics, marketing, sales and distribution network. Standard Textile provides utmost satisfaction to customers from all over the globe in a wide variety of textile and apparel products. Its operations include the following.

- North American Operation
- Mexican and Central American Operations
- European Operations
- Middle-East Operations
- Far-East Operations

Vintage Textile

The term 'Vintage' originally referred to the year a wine or oil was bottled. Then gradually the word passed into general usage to mean the year a particular item was made. First, it was used to describe vintage cars, meaning cars that were at least 50 years old and later on it started to include all sort of old products.

What is vintage textile?

Any textile fabric or product to be included in the category of vintage textiles must essentially belong to another era and mostly because of its limited supply that people develops a yearning to collect them. Another important aspect of vintage textile that it should possess an exclusive quality that would make it as desirable as it was in its heyday. Desirability and quality are two factors that actively contribute in making a vintage textile exclusive.

Pros and Cons of vintage textiles A vintage textile item is bound to exhibit some signs of wear and it is unrealistic as well as unreasonable to expect a vintage textile to be in the same condition as that of a new garment and to possess the same durability. Thus all vintage textile products need to be treated with post care and the wearable vintage textile garments and accessories requires gentle wearing and proper maintenance.

Common Sources of Vintage Textiles :

- Different types of auctions
- Specialist shops
- Markets
- Re-enactment sites
- Online sites of vintage

Vintage clothes are usually selected and sorted from old clothes and frequently sold through auctions arranged by various auction houses. The auction houses not only regularly sells textile garments and accessories but also includes textile products like linens, drapes, curtains, shawls etc. Vintage textile collectors can access information regarding vintage textiles through catalogues available at the auctions.

Maintaining Vintage Textiles A person needs to be cautious while dealing with a vintage textile as cleaning and washing a vintage textile needs post care.

Considerations for proper maintenance of vintage textiles

- It is advisable not to use washing machine or dryer to clean and wash any vintage item.

- One item should be washed at a time.
- While washing and cleaning it is best to use water which is diluted with one part boiling to three parts cold water.
- The vintage item should never be agitated and mildest of detergent should be used.
- The item should never be tumble dried.
- When the item is almost dry or damp dry it should be pressed carefully testing the pressing on an inside facing.
- The vintage item should always be hanged on a padded hanger.

E-Textile

What is e- textile?

Electronics textile or-textile belongs to the category of newly emerging discipline in the field of research which combines specialists in respective discipline of information technology, microsystems, materials and textiles. The emphasis of this upcoming area is aimed at developing the enabling technologies and fabrication techniques for the economical production of large area and conformable information systems which are estimated to have multiple applications in both consumer electronics and military industry.

E- Textiles are most likely to generate a considerable body of research which will have far reaching implications in our everyday life, consumer market and multiple applications. It requires remote sensing, processing and actuation. E-Textile will throw new challenges for hardware and software system developers due to its unique and individualistic requirements.

Small electronic devices like PDAs and cell phones have made significant contribution in changing our day-to-day life for the past few years. Electronics textile or e textile aims ay establishing a routine structure in textiles by using standard printed wiring board. The emphasis of e-textile will be on creating a woven textile with embedded copper wires which will provide the usual wear comfort and at the same time it will have a circuit with a few chips running at few megahertz.

Use of Electronics textile or e textile : Embedding electronics in textile garments would result in inventing new application in various fields. For e.g. in the field of medical prevention and rehabilitation, it becomes highly necessary to continuously monitor the patients' health condition by keeping sensing devices close to the body and e-textile can successfully accomplish sensor functionality.

Antique Textile

The word ***antique*** refers to something that belongs to or made in an earlier period and antique textile refers to any sort of textile product which is valuable primarily because of its age and reflects the beauty, artistry, style and period of its origin.

Sources of antique textile :

- Different types of auctions
- Specialized antique shops
- Fairs for antique goods
- Online sites for antique textiles
- Markets

Proper Care for antique textiles : Antique textiles needs proper handling and good care as they tend to get damaged easily.

Handling : *Antique textiles* are usually fragile and delicate so they need cautious handling.

- Jewellery and other accessories that can cause harm to an antique item should be carefully removed.
- White cotton gloves can be used while handling such items.
- It is better to carry an antique textile costume in a white sheet across the fore arms.

Cleaning :

- An **antique textile** should never be cleaned in the same way an ordinary textile is usually cleaned.
- It is advisable to use a vacuum cleaner on a very low power setting.
- Microfilament net can be used to cover flat sturdy textiles while cleaning them.
- A textile conservator should be consulted for even the slightest dilemma or confusion.
- Boiling water should never be used while cleaning and washing antique textiles.

Storage : Specialized storage solutions can be used for protection of antique textiles.

Antique Textile Fair: The Textile Society of U.K.organizes a textile fair every year where traders from various countries offers an excellent array of antique textiles for sale. Anybody who is even remotely interested in antique textiles, this fair provides a great opportunity to collect all sort of antique textile products ranging from costumes, accessories to furnishing products.

African textiles The weavers, dyers and other textile artists of Africa together makes an active contribution in creating exquisite and amazing range of textiles. African textiles usually embody a great variety of styles. Adinkara, kente and bogolan are some of the some of the African textiles which are becoming increasingly popular while some others like Yoruba, ase-oke and adire are equally beautiful but less well known.

European textiles European countries particularly England, France and Italy occupy a significant position in textile sector. Edmund Cartwright, an English clergyman patented the first power loom in 1785and a French inventor; Joseph- Mario Jacquard introduced the mechanism of supporting the woven patterned textile creation in 1801.

Textile World This term also refers to the name of the North America's leading textile publication which covers and analyzes detail information regarding all the aspects of textile industries from fibre to finished product.

Knoll Textile

Florence Knoll established Knoll Textiles in 1947.Knoll Textiles is one of the largest textile suppliers of North America. Knoll Textiles offer a wide range of technically advanced materials for furnishing office and home which intermingles beauty and utility in contemporary trends and style. Dorothy Cosonas provides creative direction to Knoll Textiles and it coordinates and collaborates with cutting edge designers and artists to explore a new range of innovative materials.

Knoll Textiles is acknowledged all over the world for creating excellent quality of furnishings which not only inspires and evolves but endures as well.

Products of Knoll Textiles Knoll Textiles is responsible for creating a diverse and comprehensive range of superior quality of furniture products, textiles and accessories that provides both choice and value to the design community and their potential customers. These products are environment friendly, recommendable for health and safe to handle. To enliven and reconceive the work place knoll textiles since its inception has worked with some of the finest designers. Knoll Textiles offers a delightful and surprising collection of textile fabrics aimed specifically for the using in corporate, office, hospitality, retail, healthcare and residential spaces.

Knoll Textiles Knoll Textiles enjoys partnership with varied range of industrial organizations throughout North America.

- American Institute of Architects (AIA)
- American Society Of Designers (ASID)
- Corporate Real Estate Network (CoreNet)
- CREW Network
- International facilities Management Association (IFMA)
- International Interior Design Association (IIDA)
- U.S. Green Building Council (USGBC)

The Knoll Design Heritage: It strongly believes that furniture should be capable of celebrating function, form and style and should be manufactured in such a way that it can weather the test of time.

Knoli Workplace Research: Independent field studies; collaborations with customers, universities and industry associations; and ongoing dialogue with experts from the fields of technology, architecture, design, organizational behavior, psychology and business management comprise the wide range of research activities in knoll Textile.

The products, policies and practices Knoll textile is designed to fulfill the purpose of protecting biosphere, conserve natural resources and reduce waste.

Kitchen Textile

Textile originally referred to a woven fabric however, gradually the term textile as well as the plural textiles implies fibres, filaments and yarns. Textile products are usually broadly categorized as textile apparel and textile home furnishing.

Kitchen textiles fall into the category of home textiles or home furnishing textiles and almost all the fabrics used in home furnishing are utilized in kitchen textile products to keep up with modern trends.

A wide range of innovative products are available in the category of kitchen textiles which cater to the varying requirements and budgets of consumers. Kitchen textiles comprise of the products and the fabrics used to produce kitchen products. Products which are primarily used in our kitchen both for functional purposes as well as decorative purposes are known as kitchen textile products.

The **textile fabrics** most commonly used in are listed below.

- Silk fabric
- Cotton fabric
- Jute fabric
- Rayon fabric
- Wool fabric
- Nylon fabric
- Polyester fabric
- Plastic fabric
- Leather fabric
- Satin fabric
- Organza fabric
- Organdy fabric

Kitchen Textile Products :

- Poly bag holders
- Aprons
- Kitchen curtains
- Kitchen towels
- Kitchen wears
- Kitchen linen
- Kitchen table cloth
- Food packaging materials
- Kitchen mats
- Pot holder

- Tea cozy
- Napkins
- Wash cloth
- Kitchen rugs

Leading Indian manufacturer and exporter of textile products including kitchen products -

- Suntex

Market Place Textile

Market place textile refers to the wide variety of textile fabric and textile products available in the market place. Market textile comprise of a wide range of textile products which are in vogue according to the recent trends and fashion and market place textiles are often the most popular textiles .

The contemporary market place textile fabrics comprise of the following :

- Silk
- Cotton
- Jute
- Rayon
- Wool
- Nylon
- Polyester
- Plastic
- Leather
- Satin
- Organza
- Organdy

These fabrics are used to create a variety of trendy, traditional as well as functional textile products.

However, Textile Market place can also refer to a popular branded site which was primarily established in 1999 to serve the textile professionals to source trade leads, advertise their companies, sell commodity items, and experience a whole range of other interactive applications. Textile Market place was founded by New Media Solutions in co-operation with the American Textile Manufacturing Institute

(ATMI). Textile Market place enjoys the attention of the entire textile industry as a trust worthy online exchange.

Textile Market place is one of the most popular and largest online information exchanges for the entire textile industry and presently it has more than 6000 members and 2400 business listings. Facilities or services provided by the Textile Market place.

- On line live auctions
- Provides online guidance to buyers
- Showcasing products
- Provides round the clock opportunity to promote and sell services of registered companies through an active store front
- Direct contact facility through spot market
- News room which accommodates thousands of articles having textile news
- Easy access and availability of wide variety of web resources which are associated with the textile industry and a variety of informative news links

Technical Textile

The Textile Institute defines technical textiles as the 'textile materials and products manufactured primarily for their technical and performance properties rather than their aesthetic or decorative characteristics '.

Technical Textiles has never been a single consistent industrial sector and market segment as it is continuously developing in different directions with varying speed and varying degree of success. The development of emerging geographical markets in China (Chinese textile) and other global industrial regions actively contributes in influencing the rapid growth of technical textile manufacturing in this century.

The term 'Industrial textile was widely used for years to refers to any sort of textile product other than apparels, household or furnishing products. With the increasing textile application in the diverging fields of medicine, hygiene, sports, transport, agriculture and other non industrial purposes. Industrial textiles are categorized as a sub group of technical textiles and it specially comprises of those particular textile products which are utilized in the operational procedure of manufacturing.

Technical textiles are usually applied in the following 12 main areas :

- Agrotech comprises of agriculture, horticulture, aquaculture and forestry
- Buildtech comprises of building and construction.
- Clothtech comprises of technical components of footwear and clothing.
- Geotech comprises of technical components of geotextiles and civil engineering.
- Hometech comprises of technical components of furniture, household textiles and floor coverings.
- Indutech comprises of filtration, conveying, cleaning and other industrial uses.
- Medtech comprises of medical field and hygiene.
- Mobiltech comprises of automobiles, shipping, railways and aerospace.
- Oekotech comprises of environmental protection.
- Packtech comprises of packaging.
- Protech comprises of personal and property protection.
- Sporttech comprises of sports and leisure.

Technical textiles in various contexts are often referred to as performance textiles or functional textiles or engineered textiles or high-tech textiles.

Technical Textile fibre: Almost 90 % of the fibres utilized to make technical textiles belong to the category of conventional fibres and the fibres specially developed to specifically use in technical textiles are not only expensive but often have limited applications.

Technical Textile products: Within the composites industry, woven, knitted braided, non woven and wound reinforcements produced from glass, carbon fibre and organic polymer materials like aramids are presently accepted as technical textile products.

7

Clothing

Clothing or **clothes** protect the human body from extreme weather and other features of the environment. It is worn for safety, comfort, modesty and to reflect religious, cultural and social meaning. Clothes today, in the 21st century are often worn more often for fashion purposes rather than warmth and comfort.

The practical function of clothing is to protect the human body from dangers in the environment: weather (strong sunlight, extreme heat or cold, and precipitation, for example), insects, noxious chemicals, weapons, and contact with abrasive substances, and other hazards. Clothing can protect against many things that might injure the naked human body. In some cases, clothing protects the environment from the clothing wearer as well (example: medical scrubs).

Humans have shown extreme inventiveness in devising clothing solutions to practical problems and the distinction between clothing and other protective equipment is not always clear-cut; examples include spacesuit, air conditioned clothing, armor, diving suit, swimsuit, bee-keeper's protective clothing, motorcycle leathers, high-visibility clothing, and protective clothing.

People also decorate their bodies with makeup or cosmetics, scented perfume, and other ornamentation; they also cut, dye, and arrange the hair on their heads, faces, and bodies (see hairstyle), and sometimes also mark their skin (by tattoos, scarifications, and piercings). All these decorations may contribute to the overall effect and message of clothing, but do not constitute clothing.

Articles carried rather than worn (such as purses, canes, and umbrellas) are normally considered fashion accessories rather than clothing, but hats and small dress sweaters can be called clothing or accessories. Jewelry and eyeglasses are usually considered as accessories as well, even though in common speech these particular items are described as being worn rather than carried.

Historically, there have been many societies where partial or complete nudity was the cultural norm. In the modern world, most cultures find it socially unacceptable to walk around in public without clothes.

History and Origin of Clothing and Textile

The History of clothing and textiles attempts an objective survey of clothing and textiles throughout human history, identifying materials, tools, techniques, and influences, and the cultural significance of these items to the people who used them.

Textiles, defined as felt or spun fibres made into yarn and subsequently netted, looped, knitted or woven to make fabrics, appeared in the Middle East during the late stone age. From ancient times to the present day, methods of textile production have continually evolved, and the choices of textiles available have influenced how people carried their possessions, clothed themselves, and decorated their surroundings.

Sources available for the study of the history of clothing and textiles include material remains discovered via archaeology; representation of textiles and their manufacture in art; and documents concerning the manufacture, acquisition, use, and trade of fabrics, tools, and finished garments. Scholarship of textile history, especially its earlier stages, is part of material culture studies.

Some human cultures, such as the various peoples of the Arctic Circle, until recently made their clothing entirely of prepared and decorated furs and skins. Other cultures have supplemented or replaced leather and skins with cloth: woven, knitted, or twined from various animal and vegetable fibres.

• **Weaving** is the textile art in which two distinct sets of yarns or threads, called the warp and the filling or weft (older *woof*), are interlaced with each other to form a fabric or cloth. The warp threads run lengthways of the piece of cloth, and the weft runs across from side to side.

Warp and weft in plain weaving.

Cloth is woven on a loom, a device for holding the warp threads in place while the filling threads are woven through them. *Weft* is an old English word meaning "that which is woven".

The manner in which the warp and filling threads interlace with each other is known as the **weave**. The three basic weaves are plain weave, satin weave, and twill, and the majority of woven products are created with one of these weaves.

Woven cloth can be plain (in one colour or a simple pattern), or it can be woven in decorative or artistic designs, including tapestries. Fabric in which the warp and/or weft is tie-dyed before weaving is called ikat. Fabric decorated using a wax resist method is called Batik.

The ancient art of handweaving, along with hand spinning, remains a popular craft. The majority of commercial fabrics in the West are woven on computer-controlled Jacquard looms. In the past, simpler fabrics were woven on dobby looms, while the Jacquard harness adaptation was reserved for more complex patterns. Some believe the efficiency of the Jacquard loom, with its Jacquard weaving process, makes it more economical for mills to use them to weave all of their fabrics, regardless of the complexity of the design.

- **Knitting** is a method by which thread or yarn may be turned into cloth. Knitting consists of loops called stitches pulled through each other. The active stitches are held on a needle until another loop can be passed through them.

Knitting may be done by hand or by machine. By hand, there are numerous styles and methods. Flat knitting, which can be done on two straight needles or a circular needle, produces a length of cloth, while circular knitting, which is done on circular or double-pointed needles, produces a seamless tube.

Different yarns and knitting needles may be used to achieve different end products by giving the final piece a different colour, texture, weight, or integrity. Using needles of varying sharpness and thickness as well as different varieties of yarn adds to the effect.

Another approach involves cutting and sewing the cloth, but using every bit of the cloth rectangle in constructing the clothing. The tailor may cut triangular pieces from one corner of the cloth, and then add them elsewhere as gussets. Traditional European patterns for men's shirts and women's chemises take this approach.

Modern European fashion treats cloth much more prodigally, typically cutting in such a way as to leave various odd-shaped cloth remnants. Industrial sewing operations sell these as waste; home sewers may turn them into quilts.

In the thousands of years that humans have spent constructing clothing, they have created an astonishing array of styles, many of which we can reconstruct from surviving garments, photos, paintings, mosaics, etc., as well as from written descriptions.

Social Status

In many societies, people of high rank reserve special items of clothing or decoration for themselves as symbols of their social status.

In ancient times, only Roman senators were permitted to wear garments dyed with Tyrian purple; only high-ranking Hawaiian chiefs wore feather cloaks and palaoa or carved whale teeth. Under the Travancore kingdom of Kerala (India), lower caste women had to pay a tax for the right to cover their upper body. In China before the establishment of the republic, only the emperor could wear yellow. In many cases throughout history, there have been elaborate systems of sumptuary laws regulating who could wear what. In other societies (including most modern societies) no laws prohibit lower-status people wearing high status garments, but the high cost of status garments effectively limits their purchase and display. In current Western society, only the rich can afford haute couture. The threat of social ostracism may also limit garment choice. If one is not wearing a specific brand or style of clothing one's social status may fall. Yet, retailers have converted to fast fashion techniques that provide trendy pieces of clothing at lower price points.

Religious habits and special religious clothing

Religious clothing might be considered a special case of occupational clothing. Sometimes it is worn only during the performance of religious ceremonies. However, it may also be worn everyday as a marker for special religious status.

For example, Jains wear unstitched cloth pieces when performing religious ceremonies. The unstitched cloth signifies unified and complete devotion to the task at hand, with no digression.Sikhs wear a turban as it is a part of their religion.

The cleanliness of religious dresses in Eastern Religions like Hinduism,Sikhism, Buddhism and Jainism is of paramount importance, which indicates purity.

Clothing figures in prominently in the Bible where it appears in numerous contexts, the more prominent ones being: the story of Adam and Eve, Joseph's cloak, Judah and Tamar, Mordechai and Esther. Furthermore the priests officiating in the Temple had very specific garments, the lack of which would make one liable to death.

Sport and activity

Most sports and physical activities are practiced wearing special clothing, for practical, comfort or safety reasons. Common sportswear garments include short pants, T-shirts, tennis shirts, tracksuits, and trainers. Specialized garments include wet suits (for swimming, diving or surfing), salopettes (for skiing) and leotards (for gymnastics). Also, spandex materials are often used as base layers to soak up sweat. Spandex is also preferable for active sports that require form fitting garments, such as wrestling, track & field, dance, gymnastics and swimming.

Clothing materials : Common clothing materials include natural fibres, which are renewable, biodegradable, such as:

- Cloth, typically made of viscose cotton, flax, wool, ramie, silk
- Down for down-filled parkas
- Fur
- Leather
- Denim

And synthetic fibres which are man made and not biodegradable, made primarily from petrochemicals:

- Spandex
- Polyester
- Nylon

Less-common clothing materials include:

- Bamboo
- Hemp
- Jute
- Lyocell
- Recycled paper
- PVC-Polyvinyl chloride
- Rayon
- Rubber
- Recycled or Recovered Cotton
- Recycled PET
- Soy
- Tyvek
- Other Natural Fibres

Reinforcing materials such as wood, bone, plastic and metal may be used in fasteners or to stiffen garments.

Clothing maintenance : Clothing suffers assault both from within and without. The human body sheds skin cells and body oils, and exudes sweat, urine, and feces. From the outside, sun damage, moisture, abrasion and dirt assault garments. Fleas and lice may hide in seams. Worn clothing, if not cleaned and refurbished, will itch, look scruffy, and lose functionality (as when buttons fall off and zippers fail).

In some cases, people wear an item of clothing until it falls apart. Cleaning leather presents difficulties, and bark cloth (tapa) cannot be washed without dissolving it. Owners may patch tears and rips, and brush off surface dirt, but old leather and bark clothing will always look *old*.

But most clothing consists of cloth, and most cloth can be laundered and mended (patching, darning, but compare felt).

Humans have developed many specialized methods for laundering, ranging from the earliest "pound clothes against rocks in running stream" to the latest in electronic washing machines and dry cleaning (dissolving dirt in solvents other than water).

Many kinds of clothing are designed to be ironed before they are worn to remove wrinkles. Most modern formal and semi-formal clothing is in this category (for example, dress shirts and suits). Ironed clothes are believed to look clean, fresh, and neat. Much contemporary casual clothing is made of knit materials that do not readily wrinkle, and do not require ironing. Some clothing is permanent press, having been treated with a coating (such as polytetrafluoroethylene) that suppresses wrinkles and creates a smooth appearance without ironing.

Once clothes have been laundered and possibly ironed, they are usually hung on clothes hangers or folded, to keep them fresh until they are worn. Clothes are folded to allow them to be stored compactly, to prevent creasing, to preserve creases or to present them in a more pleasing manner, for instance when they are put on sale in stores.

Many kinds of clothes are folded before they are put in suitcases as preparation for travel. Other clothes, such as suits, may be hung up in special garment bags, or rolled rather than folded.

Many people use their clothing as packing material around fragile items that might otherwise break in transit.

Mending : In past times, mending was an art. A meticulous tailor or seamstress could mend rips with thread raveled from hems and seam edges so skillfully that the darn was practically invisible. When the raw material — cloth was worth more than labor,. it made sense to expend labor in saving it. Today clothing is considered a consumable item. Mass-manufactured clothing is less expensive than the labor required to repair it. Many people will buy a new piece of clothing rather than expend time mending. The thrifty still replace zippers and buttons and sew up ripped hems.

The life cycle of clothing : Used, unwearable clothing was once used for quilts, rag, rugs, bandages, and many other household uses. It could also be recycled into paper. Now it is usually thrown away. Used but still wearable clothing can be sold at consignment shops, flea markets, online auction, or donated to charity. Charities usually skim the best of the clothing to sell in their own thrift stores and sell the rest to merchants, who bale it up and ship it to Third World countries, where vendors bid for the bales, then sell the used clothing.

There are many concerns about the life cycle of synthetics which come primarily from petrochemicals. Unlike natural fibres, their source is not renewable (in less than millions of years) and they are not biodegradable.

Evaluation of the Properties and Clothing Comfort

One of the main purposes of clothing is to provide the wearer protection against undesirable environments. The properties of clothing materials have critical influences on the comfort of the wearer. Also, clothing is not just a passive cover for the skin, it interacts with and modifies the heat regulating function of the skin and has effects that are modified by the environment condition; Up to the present, most physiological studies have been on the thermal regulation of the human body without clothing. Although it is a necessary first step, more realistic and valuable information can only be obtained through studies of the interaction between clothing and the physiological aspects of the human wearer. This study reports an investigation into the combined effects of the properties of the clothing material and environment conditions on the rectal temperature of human wearers. The rectal temperature was the highest for the clothing with the lower air permeability and moisture regain during both the cooling and heating periods. In the hot environment after heating, the rectal temperature was the lowest for the clothing with the higher air permeability and moisture regain in environments of both with and without wind.

Wear trials of a woven and a knitted Scandinavian cross-country ski dress with identical underwear and other garments were performed in a climatic chamber on 5 subjects who walked at -9°C on a treadmill with a 5% grade for a total of 2 hr, changing the velocity of the walk every 20 minutes from 4.1 to 6.7 km/hr. The energy output corresponded to 200, and 350 kcal/m^{2}hr, respectively.

One-half of the produced sweat did not evaporate during the experiments and remained trapped in the clothing. Though the dress proper and its underwear covered two-thirds of the body surface area (trunk, arms, thighs), more than one-half of the not evaporated sweat was found in the garments of the peripheral area (head, hands, legs, feet). The vest contained a high amount of water, and the same partial pressure of water vapor was recorded in the spaces on either side of it, indicating that the beginning of the effective diffusion gradient through the clothing was shifted from the skin to the outer side of the vest. In the absence of wind in the trials, the clinging knitted dress assembly had a higher thermal resistance than the loosely cut woven one. But, subjectively, the woven assembly was felt to be warmer than the knitted which contained more trapped sweat.

The rectal and skin temperature, the temperature and humidity in the spaces between the clothing, and the subjective thermal and humidity sensations reflected well, by regular oscillations, the cyclic changes in the intensity of metabolism. The general thermal sensations correlated better with the weighted average of the rectal and mean skin temperature than with either temperature alone. However, in the quick walking periods, the general thermal sensations were the same as the sensations in the peripheral body area, while the rating of the sensations in the central area under the dress proper was distinctly lower. Special attention should be paid to the role of the peripheral parts of the body in the evaluation of clothing comfort.

Findings – The results show how business clothing systems influence on the dry and evaporative heat loss between the manikin surface and environment in different environmental and sweating conditions.

Practical implications – When using sweating thermal manikin Coppelius, Water Vapour Transmission (WVT) through and water condensation on the clothing can be determined simultaneously with the thermal insulation of clothing system. Measured thermal comfort properties of clothing systems evaluated with a sweating thermal mank in can provide valuable information for the clothing industry by manufacturing/designing new clothing systems.

Originality/value – In this investigation, the heat and moisture transmission properties of male business clothing systems were measured in different environmental and sweating conditions. In the past few years, clothing materials containing microencapsulated Phase-Change Materials (PCMs) have appeared in outdoor garments, particularly sportswear; therefore, we decided to investigate the thermal comfort properties of different standard male business apparel, as well as male business clothing that contain PCMs used as liner and outerwear material.

The thermal properties of clothing are essential in most types of human activities. Although the human thermo-regulation limits are fairly small, thermal comfort can be maintained in a very wide variety of environment and activity conditions by a proper selection of clothing. The two properties that are of primary importance are thermal insulation (resistance to dry heat loss) and water vapour permeability (resistance to evaporative heat loss). Particularly for outdoor use, air permeability and resistance to water penetration are additional important factors.

In northern Europe, exposure to cold temperatures is most commonly associated with outdoor activities in the wintertime, whereas in central and south Europe the problem is more associated with work in artificially cooled cold stores. The health risks from exposure to cold depend on several factors: temperature, wind, activity, duration, protective clothing, personal fitness and acclimatisation. Effects are associated with general as well as local cooling of for examples hands and feet. Overall adequate clothing facilitates the warming of hands and feet, whereas inadequate protection of the extremities interferes with the overall heat balance. General body cooling affects behaviour and work capacity, and may ultimately result in hypothermia. A hypothermic person is extremely susceptible to further exposure and progressive body cooling. External help is required for the alleviation of cold stress. General and local cooling makes a person more prone to misjudge information and make mistakes. In a cold state the body may be more prone to infections.

Clothing According to Different Conditions

Clothing must be adapted to the climate in which a person lives. Warm or heavy clothing is rendered imperative in a northern climate, while the lightest and thinnest can only be tolerated in the torrid zones. It is, however, a physiological fact that the more the whole surface of the body is exposed to the external air, within certain limits, the more vigorously is its functional action performed, and the better is it enabled to preserve its own proper temperature, as well as to resist all unwholesome impressions from vicissitudes of weather, or the extremes of heat and cold. It should always be as light and loose as possible without bodily discomfort.

The substances principally employed for clothing are linen, cotton, silk, wool, hair, or down. Woollens or flannels, being bad [illegible] of heat, afford the greatest immediate protection from cold; and for the same reason are less debilitating to the cutaneous function than is generally supposed. The healthiest clothing for a cold climate, especially the year round, is undoubtedly that made of wool. If worn next to the skin by all classes in summer and winter, an incalculable amount of coughs colds, diarrheas, dysenteries, and fevers would be prevented, as also many sudden and premature deaths from croup, diphtheria, and inflammation of the lungs and bladder. Of course, the clothing should be regulated in amount according to the degree of the heat of the weather at the time prevailing. In a very hot day, for instance, a single garment might be sufficient, but on a colder day an additional garment should be added and in this way keep the equilibrium of the temperature of the body uniform as possible day by day, the year round. Winter maladies would be prevented by the ability of a woollen garment to keep the natural heat about the body, instead of conveying it away as fast as generated, as is done by linen, flaxen, cotton, and silken garments. Indeed, the laboring classes, or those compelled to toil in the sun, would enjoy better health by wearing light woollen clothing, than by wearing linen or cotton fabrics. Among the Irish emigrants and others who arrive in the United States during the summer season, we find many clothed entirely in woollen garments, frequently wearing heavy cloaks or coats, and actually feeling less discomfort from the heat than those of our native-born citizens who are in the habit of wearing linen or cotton next to their skin, and similar fabrics over these for outer clothing. It is more healthful to wear woollen next to the skin, especially in summer, for the reason that woollen textures absorb the moisture of perspiration so rapidly as to keep the skin measurably dry all the time. It is curious to notice that the water is conveyed by a woollen garment from the surface of the body to the outer side of the garment, where the microscope shows it condensed in millions of pearly drops; while it is in the experience of all observant people, that if a linen shirt becomes damp by perspiration, it remains cold and clammy for a long time afterwards, and, unless removed at once, will certainly cause some bodily ailment, as palsy, rheumatism, etc. To sit down, or remain inactive with linen or cotton shirt wet with perspiration, will speedily cause a chill to the whole body, leading not unfrequently to some sudden and fatal disease. In the night-sweats of consumption, especially, or of any debilitated condition of the system a woollen or flannel night-dress (light for warm weather) is immeasurably more comfortable than cotton or linen, because it prevents that sepulchral dampness and chilliness of feeling which are otherwise inevitable. The British government makes it imperative that every sailor in the navy shall wear flannel shirts in the hottest climates, a rule that should be adopted by all persons everywhere exposed to variable weather, to extreme heats and colds, merely regulating the amount of woollen garments worn to suit the variable temperatures of climates and seasons. In saying all this, however, we must remember that comfort is very much a matter of habit; and therefore we should make due discrimination between the natural sensation of health and the morbid sensitiveness produced by false customs. For instance, some keep their whole bodies constantly covered by many layers of woollen garments, and yet go into a shivering fit at every unusual breath of cold air. The reason is, they never adapt their habiliments gradually to the degree of the heat or cold of the season. If it is to be deemed advisable to wear woollen clothing all the year round; whether summer or winter, it does not follow that we are to wear more than one or two extra folds of clothing in addition to the under garments. The true rule is not to cover all parts of the body equally with the same amount of clothing. The fleshy parts require the least clothing, and the limbs and feet, of less muscular parts, the most. Yet we often wear, in addition to under-clothing, a thick vest, coat, and overcoat; and to these will add heavy scarves of fur or wool to the neck, etc., while the legs and feet are seldom clad in more than a single additional garment to the drawers and stockings. These parts require more clothing, especially in the winter season, than any other parts of the body. Furs are worn in the United States more for ornament than benefit. They are the warmest clothing materials known; yet are not adapted for general wear, in as much as they are apt to overheat the body, and thus render it keenly susceptible to colds and other afflictions. By consequence, fur neck cloths, caps, etc, are very pernicious for the head and throat, inducing catarhs, quinsy sore throat, and similar afflictions. On the contrary, a light woollen waist-coat worn constantly over the breast, summer and winter, would guard against these and other evils, and insure vigorous strength to the lungs or respiratory apparatus, and thus should not be dispensed with even in dog-days. The simple rule is to keep the head cool and the feet warm at all seasons of the year. Cheap and pretty silks, of which there are many varieties, are materials which are admirable for ladies' evening, dinner, or walking dresses, and cost less in the end than other fabrics.

In a strictly hygienic regulation of dress, however, the colour of the clothing is not to be disregarded. White colour reflects the rays of the sun; black absorbs them. Light coloured clothing is, therefore, more comfortable and sanitary in warm weather than dark coloured, because the former repels the heat, while it is readily received and retained by the latter. The heat-reflecting or heat-retaining property of different fabrics varies exactly with their lighter or darker shades of colour. This difference, however, is much greater in the luminous rays of light than in the non-luminous. When, therefore, we are not exposed to the sun, the subject of colour is of very little importance. The absorbing power of dark surfaces renders the skins of dark-coloured animals, as well as the darker persons or races of the human family, less liable to be scorched or blistered by the direct rays of the sun than are those of a lighter colour.

As to the cut or fashion of garments, that is a matter to be decided by the taste or habits of the wearer. Fashion, however, is very arbitrary, and seldom consults hygiene in matters of dress. Of late years she has really much improved, as to the regulation of attire with regard to both health and elegance. The hooped skirt, which at the outset of its career was so mercilessly ridiculed, has proved to be a great blessing to the ladies, as it enables them to dispense with a heavy drag of solid skirts, and gives their lower limbs free and easy play and motion. The hat or head-coverings now worn by both sexes are, in a sanitary point of view, far superior to those worn by our immediate ancestors, being very light, and affording free ventilation, which is indispensable for the avoidance of headaches, rushing of blood to the head, and many other afflictions.

Inequality of clothing, as before remarked, is a far more frequent cause of colds than deficient clothing. For instance, if a person exposes a part of the body usually protected by clothing to a strong current of cold air, he will take cold sooner than by an equal exposure of the whole body. A great safeguard against disease is to regulate the texture and quantity of clothing according to the temperature of the climate in which a person lives, avoiding extreme colds or extreme heats; keeping the clothing always fresh and clea (especially that of the feet), and wearing a different garment night from that worn during the day, not omitting the cleanliness the whole body in the general hygiene of wearing apparel.

Clothing: Constitutions and Habits

The clothing ought always to be proportioned to th constitution and habits. A person of strong constitution, who take much exercise, needs less clothing than one of delicate and sedentary habits. According to this rule, women need much thicker and warmer clothing, when they go out, than men. But how different are our customs, from what sound wisdom dictates! Women go out with thin stockings and thin shoes and open necks, when men are protected by thick woolen hose and boots, and their whole body encased in many folds of flannel and broadcloth.

Flannel, worn next the skin, is useful, for several reasons. It is a bad conductor of heat, so that it protects the body from sudden chills when in a state of perspiration. It also produces a kind of friction on the skin, which aids it in its functions, while its texture, being loose, enables it to receive and retain much matter, thrown off from the body, that would otherwise accumulate on its surface. This is the reason why medical men direct that young children wear flannel next the body, and woolen hose, the first two years of life. They are thus protected from sudden exposures. This is the reason, too, why labouring men should wear flannel next the body. Flannels are also considered as preservatives from infection, in unhealthy atmospheres. The give a healthy action to the skin, and thus enable it to resist the operation of unhealthy miasma. For this reason, persons residing in a new country should wear flannel next the skin, to guard them from the noxious miasmas caused by extensive vegetable decompositions.

But the practice by which females probably suffer most, is the use of tight dresses. Much has been said against the use of corsets by ladies. But these may be worn with perfect safety, and be left off, and still the injury they often produce be equally felt. It is the constriction of dress that is to be feared, and not any particular article that produces it. A frock, or a bet, may be so tight, as to be even worse than a corset, which would more equally divide the compression.

Such female figures as our print-shops present, are made, not by the hand of the Author of all grace and beauty, but by the murderous contrivances of the corset-shop; and the more a woman learns the true rules of grace and beauty for the female form, the more her taste will revolt from such ridiculous distortions. The folly of the Chinese belle, who totters on two useless deformities, is nothing, compared to that of the American belle, who impedes all the internal organs in the discharge of their functions, that she may have a slender waist.

It was shown, in the article on the bones and muscles that exercise was indispensable to their growth and strength. If any muscles are left unemployed, they diminish in size and strength. The girding of tight dresses operates thus on the muscles of the body. It an article, like corsets, is made to hold up the body, and then those muscles which are designed for this purpose are released from duty, and grow weak; so that, after this has been continued for some time, leaving off the unnatural support produces a feeling of weakness. Thus a person will complain of feeling so weak and unsupported, without corsets, as to be uncomfortable. This is entirely owing to the disuse of those muscles which corsets throw out of employ.

Another, effect of tight dress, is to stop the office of the lungs. Unless the chest can expand, fully, and with perfect ease, a portion of the lungs are not filled with air, and thus the full purification of the blood is prevented. This movement of the lungs, when they are fully inflated, increases the peristaltic movement of the stomach and bowels, and promotes digestion; and any constriction of the waist tends to impede this important operation, and indigestion, with all its attendant evils is often the result.

The rule of safety, in regard to the tightness of dress, is this. Every person should be dressed so loosely, that "when sitting in the posture used in sewing, reading, or study. 'The Lungs' can be as fully and as easily inflated, as they are without clothing." Many a woman thinks she dresses loosely, because when she stands up, her clothing does not confine her chest. This is not a fair test. I is in the position most used when engaged in common employments, that we are to judge of the constriction of dress. Let every woman then bear in mind, that just so long as her dress and position oppose any resistance to the chest, in just such proportion her blood in unpurified, and her vital organs are endangered.

8

Fabric Selection

In the selection of a fabric the consumer desires that the article should furnish him with satisfaction both immediate and remote. The three qualities demanded of a fabric are: attractiveness, suitability and durability. In other words the article should meet acceptably the requirement for which it is purchased. This needs an intelligent analysis of the situation by the consumer covering the kind of fabric i.e. velvette. The durability of the fabric is detemlined by such service qualities as washability, colour fastness and crease resistance. On the ground of such an analysis the consumer is faced with the requirement of a wide knowledge of textiles if he is to buy intelligently.

Basis for the Choice of Clothing Fabrics

The selection of clothing fabric is one of the most not easy problems of the consumer. This is true with regard to clothing for men and boys as well as that for women and children. The clothing selecting by a person living only 50 years ago would comprise garments made from fabrics of standard construction as muslin, flannel and taffeta. Selection includes no request as to whether knitted or woven materials would serve best. It was a matter of knowing whether one wished wool, cotton, silk, or linen. In the purchase of any article the amount of money to be expended is one of the governing factors in the selection. If we can spend but a few rupees a yard for cloth we require not waste time at the other counters in proportion as the amount of money to be expended is increased as selection at a given cost becomes more complex. Fabrics utilised for underwear meet certain defiant requirements such as softness of texture, ease of laundering, fastness of colour, and durability. The width of the cloth is an potent factor in the price of any article made from it. Generally a wide fabric is more economic than a narrow one th;lt costs less per yard. The quest of survivability remains one on which the intormation required by the consumer is clearly sack.

Although the breaking strength and number of ends of cloth are commonly accepted as determining factors in its ability to understand the wear and tear of content use likewise no information is available giving the resistance to the friction of rubbing. This too would connote varying qualities. Lacking this the consumer buys in the dark. The fastness of colour is the one factor in serviceability on which the consumer may in many intense obtain sufficient confirm information.

The weave is an potent factor influencing the appearance and properties of cloth. All weaves are represented in the construction of silk fabrics, although particular weaves may some times be difficult to fmd. Plain weave furnishes strength and satin weave provides a smooth surface suitable for lining material and undergarments and a high lustre i.e. desirable in fabrics to be made into clothing for formal wear. Twill weave provides service. Pile weave provides thickness and gauze weave produces sheer.

Satins are produced from slightly twisted or untwisted yarns. The tightly twisted yarns used in crepe fabrics give the desired effect, reduce the brightness and produce a fabric that can stand hardwear and is generally guaranteed not to crush or wrinkle. Some types of hard twisted silk fabric dull in finish become shiny with wear. This should be considered when selecting fabrics made from hard twisted yarns.

Wool Fabrics

Choosing of wool fabric offers perplexing problems. We know that smooth finished materials such as serges (durable woolen fabrics) catch the dust easily than do napped flannels and broad clothes. Serge is made from twisted yam comprising the best long fibres and will stand more hardwear than fabrics made from the weaker woollen yarns.

A guaranteed old wool cloth may be anticipated to give better service than one not so labelled. If this guarantee carries the brand of a well-known manufacturer and is handled by a reliable source we could with safety purchase the fabric. If, however, we choose wool cloth carrying no label, no guarantee of quality , we must rely only upon our own knowledge to guide us with wool, perhaps the quality of the fibre is of prime importance and only a microscopic or chemical test will give this information.

Cotton Goods

In the case of white cotton goods we might realise more confident of our ability to judge quality by observing size, continuity and twist of the yarn, furnished and explicit compactness of the cloth. Cotton materials, guaranteed to retain its colour under all conditions of use, are available in the market. Some fabrics are guaranteed against shrinkge. Others' labels connote that the finish is permanent or that the fabric is wrinkle-free. These labels and guarantee are a big step in the Tight direction but fall short off giving information as. to the wearing qualities of the cloth.

The qualities of "ever fast" is a label that absolutely gaurantees the fabric against any type of fading. This label is used only on cloth of high quality so that quality is also assured.

Linen Fabrics

Linen Fabrics have limited use as clothing. Thin fine linens are suitable for children's garment and for underwear. The handkerchief is the one article for which linen is almost widely considered to be the best fi... The high absorptive property of linen and the ease with which it can be cleaned make it most requir... fabric for this use. The fact that linen crushes more readily than other materials counteracts to some extent.

Rayon and acetate

The selection of fabrics made from rayon or acetate presents even more complex problem. Some rayon cloths provide the appearance of silk and may in several examples be substituted successfully for it. The rapid development of new synthetic fibres is adding greatly to the problem of the purchaser. Nylon, vicara, arlon, dynel and dacron are among the new synthetic fibres which are now available in the market. Vicara possesses many of the desirable qualities of wool but due to its lack of strength, is mixed with wool or rayon in the construction of a fabric. These fabrics possess a warm silk like hand and can be readity laundered and dried rapidly. The fibre is particularly resistant to sunlight, moth and mildew. It is considered even better than nylon for curtains. It is recommended for clothing fabric due to its pleasing texture, and its recovery from wrinkling. Dacron possesses properties more like those of wool than any other fabric since it retains its shape, absorbs little and resists wrinkling. This fibre is very recommended for summer suitings blouses, dresses and man's shirt.

Clothes for Different Occasions

Campus Clothes

Campus clothes are both easy to make and easy to find in the ready to wear department of a store but it is not incessantly easy to meet all of your specification in one garment. People who do not sew often require to be more careful in wardrobe planning since ready to wear garments are increasing in cost and dress makers are not easy to find apart from being becoming and keyed to one's basic wardrobe colour. a school outfit should have the following chracteristics—

1. Well tailored: They look neat, hold the press.

2. Durable: Colourfast, preshrunk and hold its shape.

3. Easy to care for: They should be free of fuzzy trimmings, clearable at little expense or trouble. They should not be too tight and sexy, not baggy or sloppy looking. A generous supply of shirts plus blouses and sweaters have formed the backbone of campus wear. They add a few belts, scarves, chains, or pins for different type. Skirts, dresses and coats must be of a fabric design which can take much sitting. The material should be strong, sturdy cotton fixable nylon which is less affected by washing and perspiration.

Office Clothes

If you do work in a campus office you may wear about the same type of clothes you wear to class. They should be a little neater and more business-like--not too short, not too tight, not too thin. Keep make up and hair, dress simple.

A personnel manager asserts, "It is unfortunate that several men and women of main ability miss chances for a big job or other business success since they just don.t look the part. First impressions are based on appearance, co- ordination of tastefully chosen clothes and accessories appropriate cared for and the all important factor of grooming.

Sportswear

Every college girl should take participation in some active sport of course it is pleasant to excel in a sport, but the main objective is to enhance one's vigour and pleasure in living. It will improve your pasture develop deep breathing and afford lots of fun. Discover the most carefully clothes for each sport in your section of the country.

Dates

Code dates. dinner date. show date, library date: divide your date. For several of them you will wear ordinary campus clothes.

Special-date Clothes: Particular date clothes. however, may be more elaborate than school clothes. These are the clothes some times referred to as your 'best' or 'before five '. clothes. The designs may flare more and have draped or fancy seaming, more trimming and dainty touches. A void skimpy cuts, poor stitching, narrow seams and hems, painful trims and poor colours. is the suggestion for dressing up your more tailored garments.

Laterally

Clothes for after 5, like special dinners, may be more dressy than 'before-5' clothes. Skill in the change of accessories can make the same dress you wear in the afternoon acceptable and successful for evening.

College girls generally require two kinds of evening dress, a long or short formal for evening and an informal dress, which may have other uses. The formal kinds is more detail, low-necked and sleevesless, or with very short cap sleeves. The informal type is similar in style, material and decoration may have sleeves and has a higher neck than the formal kind. Full short skirted, party dresses with low necks and abbreviated sleeves are gay and rather sensible. When in fashion, either full length or short length is proper for formal dances.

The informal or dinner kind of dress may be needed for certain functions. A practical solution is a jacket that concerts a formal into an informal kind of costume. Most freshmen are not very successful in long tubular, black evening dresses. The average young college girl is youthful natural, and most beautiful in an evening suit designed on simple lines not too narrow, not too bouffant. A changed empire style is soften and more flattering than a plain sheaths. For a square dance or any other active dance, a full skirt order is a must-The girl with a flat bony chest or an extra full bust should not wear a strapless top, and the girl with large hips should avoid an unusually long torso style. Any snug style sheath, empire, directoire needs a well-controlled slim figure. Do not utilise net or chiffon unless you purchase enough yardage to have the skirt full. Sheer is anything but graceful. In utilising sheers, do not have the slip too light or too short. No one is attractive if the foundation garments are noticeable, a slip too makes one seem graceful.

Gloves and an evening bag that harmonize are right for a dinner dress in a city, but they are not necessary at campus and small town parties. A small hat is correct with a theatre suit or in a public dining room, but not necessary in a private dining room at a hotel with a simple evening or afternoon dress. Slipper matching some of yours date or daytime clothes are generally proper for informal evening dresses. For formal wear one pair of slippers may last for several years. If silver or gold, select whichever kind more about harmonize with your jewellery and the feasibilities with several costumes. Begin your costumes with a pair of light coloured onl.'s. later they may be dyed. Bright red is very attractive but not wearable with all colours and you may soon be tired of such a stricking colour, though today it is almost classic. White good wool is always attractive for both winter and summer but needs more repeatedly cleaning. For spring, other fabrics-cotton, linen, silk or synthetics is interesting. Textures such as cotton knitted lace are flattering and economical, particularly if you can make them yourself. A fur generally makes a young girl seem matronly, A mock fur cardigon is a versatile wrap.

If your dress is on the fancy side, keep to a simple style of hair arrangement. Study colours in a costume and make up under an artificial light before buying. Sometimes you have the opportunity of selecting for your own storage. The colour and size should be sealed to your colour ensemble and your size, consider the best placement. A fancy ornament detracts from a hair do and it makes a confusing design on a flowered or other patterned fabric like lace.

Wedding Dresses

Wedding dresses should be modest with neck and arms not too uncovered. A formal evening dress would be in poor taste. The standard bridal attire needs a head covering and gaudiness of some type, preferably or too light a dress as a wheel or flowers long sleeves or long gloves or heavy gold work with bright shade in Indian customs and a moderately high neck. On the conventional white wedding dress with long sleeves should be swag fitting at the wrist to achieve a graceful effect. If you wish to use the wedding dress as an evening dress later, there are several feasible adaptations. You might replace long sleeves with short sleeves or gold work can be limited and combined with silk threads, a shear yolk may be removed, or the dress may be finished later to des re l fitting. If you are to be married in a suit or afternoon kind of dress, the best choice is main style, wear gloves and a small hat with a modest veil. The design of the costume should be classic in simplicify and the style of hair dress and hat beautifully harmonious with the shape of your face. Photographs taken of you should be beautiful enough for your grandchildren to display proudly rather than to remark about the amusing grotesque styles of yesterday.

Simple, home, garden wedding suggests cotton suits and simple style. Bridemaids' dresses should be selected with care so that they may be worn for several future occasions and still be in good taste.

Travel

Week-end trips, summer vacation and going back and forth between home, college and for job. College girls are repeated experiences for every one. Pants may not be objectionable when driving from one town to another, but one might feel more suitably dressed in other clothings, i.e. suits. The new textiles are particularly good for long trips since they do not display soil and wrinkles. In hot, dusty summers particularly on a bus, a wash dress like cotton synthetic or a wrinkle-shed cotton is ideal. Air-cooled coaches are often chilly enough to require a jacket or sweater in mild winters.

Accessories should be durable, neat and restrained in style. Fresh gloves may be brought out to finish the journey, ours during day time should be simple on pumps or walking styles with low or medium heels. Practical for driving are T-shirts and short jackets with deep armholes for men. Packable fabrics comprise all wrinkle-resistapt fabrics, such as knits, pure silk, 100% wool and several synthetics. For tropical climates choose cottons that "breathe".

Periodicals concerned with travel today condense wardrobes to a minimum like wardrobe with 8 Dresses-"around the world in 80 ounces." Occasionally only 5 or 6 outfits are recommended for selecting a basic wardrobe is fancied since it is so sensible, not only for travel, but also for day to day living.

Curtains-Large heading

All fabrics that hang over the windows are known as curtains, no matter what their weight is. On the other hand draperies are the fabrics that frames a window.

Draperies add colour and interest and soften the lines of the doors and windows. They may be made of cotton, linen, rayon, silk, wool or mixture of these fibres. The fabric used should similar to with furnishing in the room. If the room is formal texture should lend the proper spirit. Cotton draperies look homelike and are less costly. Linens are more formal durable, serviceable, and vigrous. Silks are suggesting regalness and luxury .Synthetics are utilised in ame weaves and textures as silks.

Inner-curtains : They are hung next to window frame or glass. They may be classified in four groups—

(a) Glass curtains: Glass curtains are made of very sheer material and are hung on a rod by hooks or castings. Fibre glass, heavy lace, filet net, cotton, lawn, and cotton voile are suited for these.

(b) Tambour curtains: These can take place of glass curtains or draperies. They are made of heavy material, fairly embroidered batistes lawns, or net in detail designs.

(c) Draw curtains: Draw curtains are utilised to ensure privacy and shut out bright light. They hang from the outside rod either to the sill, or to the floor. They are made of opaque fabrics that are lined.

(d) Sash or cafe curtains: Sash or cafe curtains a. ling on a rod attached to the lower window sash or to lower half of the window casing. Any sheer fabric can be utilised for this.

Chiefly three factors are considered important while buying draperies and curtains: colour, texture and design. The colour of the curtains should not solely go well with the other furnishings but he personal taste of the owner and the climatic conditions should lso be kept in mind. Cool colours in summers and arm colours in winters should be chosen. Light colours add space to the room while dark shrinks the room. The combination of both can be utilised to give special effects or hide the excessive length and breadth of the room.

Texture combination should not be too much alike nor too much dissimilar, for instance, silk should not be combined with rough cotton. On the other hand there should be sufficient textural contrast to create interest, for instance, in a room where draperies are of silk, upholstery may be of brocade or velvet. In selection of designs it should be kept in mind that too much of pattern is tiresome and combination of plain with pattern creates interest, for instance, patterned seating arrangement goes well with a plain carpet.

The effect of sunlight on varied fibres should be kept in mind. Nearly all fabrics are weakened somewhat by sunlight. Rayon is as resistant to sunlight as cotton while acetate is more resistant than cotton. Silk is weakened in sunlight. All kinds of curtains barring gauze should be lined for protection. Linings serve as filters for sunlight, gas and dust.

Upholstery

Fabric utilised for covering upholstered furniture and cushions and slip covers is known as upholstered fabrics. It may be divided into four groups-

1. Velvets: Thick dense pile with cut loops.
2. Friezes: Looped springy fabric cross-wise row of looped uncut pile.
3. Damask: Damask (reversible) and brocades (unreversible) in elaborate jacquard designs.
4. Rib Weaves: Tapestries (with large designs), armures (with small designs) and ribs (solid column with crosswise ribs).

The fabric used for upholstery should be sturdy sufficient to resist friction, sunlight, dry-cleaning, and laundering. Silk, wool, cotton and nylon are the best for upholstery. Silk is used for luxury objective. Rayons and acetates are inexpensive and are easier to care. These are utilised in homes where there are no children who may tear or spoil them. Plastic coated fabrics are utilised for durability. Nylon is abrasion-resistant and readily cleaned. In an explicit patterns and colours can be chosen to hide soil and dirt.

The yarn should be tightly twisted and the weave should be close, to add durability. To retain the freshness of upholstery fabrics it should be covered during summer with linen or cotton. This will freshen a shabby upholstery too. The fabrics utilised for slip covers are chintz, cretonne, linen, cotton, rayon, cotton and nylon mixtures, denim and knitted stretch fabrics. Commonly utilised fabric for covering are freize, brocade, velvet, leather, tapestry, and laminated fabrics. But always the relation between room and curtains and upholstery should be kept in mind.

Floor coverings: Textile floor coverings are divided as rugs and carpets. Rugs are softer coverings laid on floor but are not fastened to it. It does not cover the whole floor. Seatter rugs are small rugs 2x3 and 4x6 ft. placed in front of doors, couches and stairs. Common size of rugs are 2 x4, 9x6, 9x10, 9x13, 9x15, 9x18, 12x12, 12x13Y2, l2xl5, l2x18, 12x21.

Carpet : Carpet is a soft covering fastened to the entire floor. Broadloom is a seamless carpeting of any weave or style. There are many decisions a consumer has to make when a floor covering is selected.

Colour: Colours harmonize if they have something in common. They may belong to same hue or colour family. Some colours are gay, light and airy whereas others are heavy and ponderous. A fragile rose cannot be introduced into a room with luxurious draperies in somber colours. This effect can be reduced by use of green or rust. The colour utilised for floor might dominate the colour-scheme but should not over power it. Seasonal changes should also be kept into consideration. Cooler places go well with woolen carpets while rugs are more suitable to marshy and warmer areas.

Design or pattern: Floor evenings may be sculptured or carved. The carved rug is made in jacquard design with various height of piles. The design depends on size of the room, style of decoration, draperies, and upholstery .Solid colour coverings show more foot falls than two-toned 'or various coloured types.

Suitability : Carpets are appropriate for coverings since they eliminate floor noises such as scraping of chairs, clicking of heels. They also absorb air borne noises and reduce slips and slides. Thus the floor coverings should be appropriate inconformity with to the home makers demand i.e. broadloom with uncut pile and tight twist for soil-resistant.

Wearing Quality : Wearing quality of floor coverings are based on-3. kind of fibres used 4. quality of the yarns 1. closeness of the weave 5. height of the pile 2. construction of the back.

Fibres: Fibres for floor coverings should have resilience, strength and length. Wool is very appropriate since it has all the properties. Cotton does not wear so well since it is less resilient and so harder to walk on. It is inexpensive and found in diverse colours, preshrunk, colourfast, small sized rugs with plastic backs can be washed readily. The price of silk makes it preventive for use in moderately priced rugs and it is not so durable. Fibre rugs are made of tightly twisted strips of paper, often vinyl coated to resist friction and moisture. They are less costly and adaptable to use in summer cabins, porches. Rugs of straw, sisal and grass are used for same purpose. Rayon and acetate, nylon and acrylic carpets are answer to limited carpet wool supply and not stable wool market.

Backings: Backings of the floor coverings are made of jute, cotton and plastic. Pre-woven backings are used for broadloom and tufted carpets.

Wool carpets are prepared from woolen or worsted yarns. Worsted yarns have little twist, are smaller in length, more glossy, more wiry and stronger than woolen yarns. They are five times more long, coarse and resilient.

Cotton yarns for rugs are made of one inch staple and are utilised for whole rug for the base of the weave and to combine surface yarns with cheaper jute yarns at back. Carpet yarns are made of single ply, two ply, three ply and four ply. It makes no difference on wearing quality whether the yam is 2 or 3 ply. As most wear is on pile yam, they must be of right size, diameter, full and lofty.

Weave: Although density of the pile is affected hy weight of the yam. The pitch and the rows of the pile per inch should also be kept in view. Pitch is the number of pile warp yarns to the 27 inch width. The greater the number closer the pile. Rows of the pile is the yarns tufts in length-wise direction, pitch is the closeness of pile surface yam. The more rows in the inch, denser the pile and greater the durability .

Height: Height of the pile is another factor of quality. Closely woven worsteds need low pile, whereas less density needs long piles. High piles are more troublesome to keep clean. A strong backing supports the pile and forms the foundation of the carpet.

Price: Tufting has speeded up the carpet construction and has brought prices down. Also cotton rugs have general price that is much lower than wool and mixtures. Nylon and rayon has a very low price and are durable and easy to keep clean.

Household Linens: The Household Linens are the textiles that serve the home. They dry dishes, hands and cover beds and tables. The division of household linens is as under—

1. Towels
2. Shower curtains
3. Bathroom fabrics
4. Table linen-a. Damask b. Embroidery c. Lace
5. Sheets and pillow-cases
6. Mattresses and pillows
7. Comforters and quilts
8. Blankets
9. Bedspreads

Towels

Bath towels are made of terry cloth with looped pile surface. It enhances the absorbency by increasing the absorptive surface of cloth. The loops should be reasonably close together, soft and not too tightly twisted. A ground work of twill weave is more durable than plain or basket weave. It keeps the loops at their place by enhancing the package. The yarns should have great tensile strength to tolerate the tension of loom and use. Poor towels are made of single thread, better greeds are made by double thread which make for durability. The selvage of the towel should be firm and even. It is important that the filling yarns wind the warps utilised for selvage. The hems at either side should be even, attentively stitched, with short constant stitches and finished at sides. The colour of the towels enhance the attractiveness of the towel. Sc,lid colours and prints can be opted. Towels come in diverse sizes. Medium size towels for bath are 20x40 inches and 22x44 inches. The largest are 24x26 and 32x64 inches. The beach-towels are even more longer and larger.

Hand towels: Hand towels seem in sizes of 16x26 and 18x36 inches, when made in colour of bath towels and wash clothes they are more useful. Face towels are made of cotton, linen and mixtures. The family rough and absorbent surface is made by yarns that are coarse, heavy, and slacky twist.

Wash clothes: The chief objective of wash cloth is to give friction for cleansing the skin. A terry cloth with IQops on both sides is better than knitted back. Knitted wash clothes are softer but stretch out of shape. Wash clothes are 12 x 12 or 13 x 13 inch.

Dish towels: Dish towels are made of cottonade, a heavy cotton correspoding to wool, crash, a rough textured cotton in plain weave and novelty yarns; damask, a cotton fabric in Jacquard pattern; glass cloth, Cotton fabric with smooth hard twisted yarns that do not lint; Osnaburg, a cotton fabric with crash like appearance; linen crash, terry cloth. High well-balanced count, tensile strength and workmanship are potent for dish towels.

Glass towels: These are intended for drying glasses and should be free of lint. Glass towelling is sold by yard.

Shower curtains

They can be opted in solid hues, stripes and figures. The fabrics of which shower curtains are made are taffeta, satin, moire taffeta, polyester and marquisette curtains over a plastic linen. The fabric utilised should be water resistant or waterproof. All plastic sheet curtains, plastic coated fabrics are waterproof. Oiled silk cloths remain waterproof as long as it lasts. Shower curtains come in sizes of 68x72 inches and 72 inches square.

Table linen

Linen is utilised widely for it looks clean, is brightness, stains can be removed readily and it washes and wears well. Table covers for indoor dining can be richer, more elegant, and quieter. While the table covers for outdoor dining can be of laminated rayon, polished cotton in solid colours, gay prints, checker boards or stripes. The damasks are made in pune linen, cotton, rayon and mixtures of these fibres. Generally there are two kinds of damasks made-simple or single, double or compound. The name applies to the kind of weave-simple damask has four float construction while double damask has seven float construction. The quality of damask relies upon the length of the fibres where the long fibres do not fuzz or pullout. The cloth should be evenly spun so that the cloth is not thick and thin in spots. The durability of damask relies on the closeness and firmness of the weave. Although the longer floats in damask give greater light reflection. The long floats are not durable.

Embroidery : Embroidery is an ornamental needle work on the fabric itself. It is done by machine or hand. Hand embroidery is more preferable and is very costly. Machine embroidery is coarser and wrong side may not be well finished. Cotton lawn, organdy, linen, silks may be embroidered.

Laces: Lace is a fabric developed by looping, interlacing, braiding and twisting threads. It is an potent trimming used on table cloths, curtains, handkerchiefs, dresses and underwear. Lace comprises of two portions that is ground which holds the patterns and another is pattern which forms closer- worked and more solid portion. Laces can be hand-made or machine-made. Linen thread is utilised for hand-made lace while nylon, Iayon, cotton and silk can be utilised for machine-made ones. The buying point of the lace relies on the fibre used for the lace and the effect of washing, bleaching, starching and ironing on the fibre.

Sheets and pillow-cases

Mojority of the bed-sheets are made of heavy cotton. Linen sheets are at easy to find. Nylon and rayon sheets are used for the silky appearance and quick drying and non-ironing properties.

While choosing the bed sheets one should keep the quality in mind. The length of the fibres should be at least 1 inch and it should be even. The Unevenness will make the cloth thick and thin at spots where thick portion will get more friction and yarns in thin portion will break and make a hole. The yarns should be spun tightly. The tensile strength of tightly spun yarns is greater than that of loosely spun yarns. The high count of yarns condition good balance of cloth. High count balance is 108 × 104 while the poor balance is

76x60. A sheet should have little or no sizing. A heavily sized bed-sheet feels flinsy and looks fuzzy after washing. A sheet should not be very light in weight or too heavy. The sheet should be 108 inches long. Smaller sheets are not serviceable and comfortable. Solid colours and prints can be opted in conformity to the choice which goes well with furnishings and mood of the room.

Pillow-cases

These are made to match the sheet. Pillowcases should be about 10 inches longer than the pillow and 2 inches larger around. Same points should be considered while buying the pillowcases as sheets. The hems should be 3 inches wide, straight and even. Seams should be firm and even.

Mattresses

The mattresses are bought for the comfort and peace. The mattresses that are too soft let the sleeper sink down deeply and does not supply good help to the body while the mattresses that are too hard do not allow perfect relaxation and often restrict circulation. Mattresses should be adjustable to diverse degrees of pressure from many parts of the body.

The filling of the mattress is largely liablt for the comfort and durability .The inner spring mattress retains its resiliency longest. Others are filled with horsehair, cotton, Kapok, foam rubber and Coir. Horse-hair mattresses are most durable and expensive. Kapok is less resilient than horse-hair. It is suitable for moist climate since it throws off moisture and dries rapidly. Cotton is less costly and it becomes lumpy with time. F oam supports all parts of body but is easy to tear and deteriorates in ultraviolet rays of sunlight.

The weave of the ticking (cover) should be in twill weave, and not in satin or sateen weave. The fibre utilised should bc cotton, rayon or mixed. Felted cotton utilised as a core of a cord stitched around the edge of the mattress makes sure flrnmess. A zipper or chain stitched seam allows removal of ticking for cleaning.

Pillows

The pillows available can be very soft in which head sinks deeply or thin and hard one. Fillings of the pillow may be graded in order of excellence starting with the best-4. Feathers 2. Kapok 1. Polyester staple fibres 3. Foam rubber.

Kapok and foam rubber are good for patients of asthama and hay fever. Kapok pillow become lumpy with time. Synthetic staple fibres are non- allergenic, buoyant, odourless, moth and mildew proof.

Blankets

Blankets may be made of 100 percent wool. Cotton, rayon, acrylic and mixtures with other synthetic fibres. Warmth of the blanket is conditioned by the thickness of the nap and not by weight. The napping by the yarns reties on the twist. Loosely twisted yarns can be napped successfully.

Cotton blankets may realise colder if used in damp climate and may shrink or felt to that extent that they feel heavy but not warm.

Blankets are made in solid colours, checks, white, plaid, plain, borders and prints. Blankets with soft texture are most attractive. The durability of blanket depends upon the length of the fibre. If the fibres are short they pullout in napping process and weaken the yam. The fibre should have good tensile strength for a long wearing. Ply yarn have greater tensile strength than single yarns. The filling yarn should be close to enhance the tensile strength of the yarn. The warp and filling count should be balanced enough with firm even weaving. Blanket when held in tight should not display thin and thick spots. The ends of the blanket should run parallel to the filling yarns. Nap of the yam should be uniform in the thicknes Heavy nap of short fibres pullout readily hence it should not be made of small fibres that are loosely woven. Bindings and finishes of the ends should be neat and strong. The edge of the blanket should be stretched with two or three rows of parallel stitching or close feather stitching. The blanket too small will be tucked in sufficiently at foot or sides and may wear out rapidly.

A blanket should be as long as length of the mattress plus the thickness of the mattress plus tucking allowance. It should have a label exhibiting size, fabric content, weight, residual shrinkage and how to care for the blanket.

9

Yarn and Fabrics Preparation

Yarns are composed of textile fibres either laid or twisted together for use in weaving, knitting or other method of constructing textile fabrics. The kind of yam chosen for a fabric affects the appearance of the fabric, its durability , hand and draping characteristics. The yarn can also increase good fibre perfomance or compensate for poor fibre performance. The structure of yarn is a fundamental factor in a finish. Fabric is a term utilised for cloth produced by knitting, weaving, felting or braiding of yarns.

Some fabric feature that are concerned to the nature of the yam, are :

(c) Texture -Smoothness, softness, crepeness.

(d) Beauty -lustre, dullness, stubbiness etc.

(a) Weight -Sheer, medium, suiting etc.

(b) Comfort -Warmth, coolness, absorbency etc.

(e) Performance -Resiliency, pilling, soiling etc.

Spinning is the process of making yarn from fibre.

Yarns can be made of-

(a) Staple fibres -Spun yarns or staple yarns.

(b) Filament fibres -Continuous filament yarns or thrown yarns. Spun yarns are prepored by the mechanical spinning of natural or man-made staple fibre. Continuous filament tow and yarn are produced by chemical spinning.

Chemical Spinning

All filament yarns barring silk are developed by chemical spinning at a fibre producing mill. Chemical Spinning is the extrusion of a solution through a spinerette. Chemical spinning comprises all of the processes from extrusion of the fibre to winding the yarns on the cone or spool for further processing or for knitting or weaving.

The number of holes in the spinerette conditions the number of filaments in the yam. The number of holes changes from 1 to 350. Filament yarns are made either with or without twist as the filament out of the spinerette If further twisting or processing needed in fabrics. the yarn is sent to a thruster

Throwing: Throwing causes all the processes utilised to put yarns in proper condition for weaving or knitting. Throwing does not beautify the yarn but lends strength and particular qualities to the yarn. Throwing basically implies twisting without drawing the filaments.

Mechanical Spinning

Mechanical spinning forms spun yarns or staple yarns in which the fibres are short staple fibres. Standard or traditional systems used in spinning staple fibres-whether natural or rnan-made into yarn-comprises the cotton system, the woollen system and the worsted system.

Newer methods of preparing yarns are open-end spinning twij'tless spinning. self-twist spinning (for ply yarns). direct spinning.faciated spinning and Bobtex system.

I. Staple Yarn

Conventional Spinning : It is the uroclean system. By the woolen and the worsted system the wool and man-made fibre can be spun into yarns. These yarns are carded, drawn and spun. Yarns made by worsted system are combed after carding to produce smoother and finer quality yarns.

Cotton System: The cotton system comprises of the series of operations performed by various machines.

The machines assist to:-

(a) Clean and parallel staple fibres.

(b) Draw them out into a fine strands

(c) Twist them to keep them together

Operations objective

Ring Spinning

This is the process of the manufacture of single yarn in the spinning operation. During spinning the roving is alternated to the desired diameter and the desired amount of twist is incoparated. The most common method of spinning is the method known as ring spinning.

The ring spinning draws, twists and winds in one regular operation. The roving is set between two sets of rollers. The front set of rollers rotates about thirty times faster than the back set. This difference in speed stretches the yarn and makes it even, smooth and uniform. The alternatedern passes through an eyelet and guided through a U-shaped guide known as a traveller, which moves around the bobbin in a ring. The movement of the traveller and the twining of the spindle on which the bobbin is held, both introduce twist into the yarn. About 12 yards of yarn is wound on to the bobbin per minute. The yarn produced thus is a single yarn.

The Woollen System: Wool and man-made fibres can be spun into yam by the woolen or the worsted system.

These yarns are carded, drawn and spun. Yarns prepared by the worsted system are combed after carding to produce smoother and fine quality yarns.

The operations of the woolen and worsted systems are :

(A) Carding and Combing.

(B) Scouring.

(C) Sorting.

(D) Spinning-Spinning of wool yam is performed by either the ring method or the mule method.

Mule Spinning: Drawing and twisting is performed in one operation and the winding in another. The machinery corers a lose of space but produces a soft yarn.

Direct Spinning

It remove the drawing but still utilises the ring spinning device for inserting the twist. The sliver is fed directly to the spinning frame. This machine is utilised to make heavier yarn for pile fabrics and carpets.

Open end Spinning

It eliminates the roving and twisting process by the ring. No knots are formed in the yarn. More yam is produced and less operator supervision is required and higher production speeds are achieved. Open end spinning leaves the step of forming the roving. The sliver is delivered by a rotatary beater.

The separated fibres are carried into the rotor by a current of air and is deposited in a v-shaped groove along the sides of the rotor. Fibres from the side of the rotor are peeled off to join the open end of a earlier formed yam that extends out of the delivery tube. As the fibres join the yam, which— is incessantly being pulled out of the delivery tube, twist from the movement of the rotor is conveyed to the fibres

Twistless Spinning

It eliminates the twisting process: A roving is wetted, drawn out, sprayed with sizing and wound on a package. The package is steamed to gelatinize. The starch binds the fibres together to form a yam.

It is utilised to make ply yarns. Yarns made from two or more yarns are twisted together. Two rovings are used at a time. Each roving is passed through a pair of rollers that are both rotating and oscillating. Each set of rollers twists its roving in a direction opposite to the other.

The roller's uneven motion forms a yam that has both twisted and untwisted areas. Adjacent ends of the two yarns are permitted to untwist around each other. Care must be taken to space the untwisted areas of each yarn in such a way that a loosely twisted section of one yam never coincides with a loosely twisted section of the other yam or a weak spot would result.

II. Fasciated Yarns : It is no longer in production due to less use and weakness of the yarn. Fasciated yarns are made from a bundle of parallel fibres held in place by surface wrapping of other fibres.

Yarns Division : Yarns can be divided into three-

I. Simple yarns.

II. Complex yarns.

III. Bulk yarns.

Simple Yarns **:** Simple yarns are the yarns, with same size and regular surface.

Simple yarn is alike in all its parts. They have diverse degree `twist rangil from loose to moderate, and light or hard twist.

Simple yarns are divided as

1. Single,

2. Ply, and

Cord.

Simple Single Yarn: Simple single yarn is suited for like weaving and knitting. A single yarn is formed by first wistin operation.

Ply Yarns are repored by second twisting operation by wo or more singles. The part of yarn is called ply. Twist is brought about by a machine in a twister. Ply yarns are twisted in opposite direction to the twist of the single. Because of this the singles wind spirally and the yarn becomes softer. It enhances the diameter, strength and durability of the yarn. In harnessing a ply yarn the number of singles utilised is written before the A wool ply, e.g. if two singles are a C used it is called two ply.

3. Cord: It is made by a. third twisting operation which.(A) Single (B) 2 ply (C) 4 ply twists ply yarns together. Some kinds of serving thread and ropes belong this group. In naming a cord it is essential to connote the number of plies and singles utilised in the cord. 4, 5 ply connote that each ply has 4 singles and 4 of these 5 ply yarns are combined to form a cord.

Characteristics of simple yarns: They produce smooth fabrics. They are durable. The uniformity of the yarns prohibit snagging and tearing.

Simple yarns are easy to maintain and have type of uses.

II. Novelty Yarns or Complex Yarns

Complex or novelty"yarns are made primarily for their appearance. Novelty yarns are made on twisters with particular attachment for giving stiff tensions to various parts which provides the yarn loose curled, twisted and looped areas. Slubs and flakes of different colours are introduced into yam by particular attachment.

Definition: Novelty yarns are defined as yarns that are irregular at regular distance. Complex yarns may be single or ply.

Novelty yarn has 3 basic parts—

(a) Ground, foundation or core. The core controls the length and stability of the end product."

(b) Fancy or effect-The Fancy or effect yarn forms the design by the way in which it is applied on core.

(c) The binder-The binder yam is utilised to attach the effect yarn so that it remains in position.

Novelty yarns are generally ply but they are not utilised to add strength to fabrics. They add interest to plain weave fabrics at a cheaper cost when its effect ply is close from the weave, their effect is generally permanent. Novelty yarns are loose and bulky and give crease resistance to a fabric but they make the fabric spongy and difficult to stitch.

Its durability relies upon the size of the novelty effect and on the firmness of the weave of the fabric. Generally smaller the novelty effect, the more durable fabric is, since the yarns are less effected by abrasions and do not tend to catch and pull out easily.

Kinds of Novelty Yarns

1. Ratine : The effect ply is twisted in a somewhat spiral arrangement around the core and at intervals a longer loop is formed which kinks back on itself and is held in place by the binder.

2. Spiral or Cork-screw Yarn: It is made by either twisting together two piles that differ in size, fibre content or diameter or by varying the rate of speed or tht, direction of twist. The spiral yarn is composed of two yarns of various size, one fine with hard twist and the bulky with slight "twist. The heavy yarn is wound spirally around the fine yarn.

The Cork-screw Yarn: The Cork-screw yarn can be prepared by twisting a fine yarn loosely and a heavy yarn so it provides the appearance of a cork- screw (It is just reverse of spiral yarn).

3. Knot, Spot, Nub or Knoop Yarn: It is made by twisting the effect ply various times in the same place two effect plies of diverse colours may be utilised and the knots are arranged to several distance. Coloured spots are received along the length of yarn. A binder is added during twisting.

Knot, Spot, Nub or Knoop yarn

4. Spike or Snarl: The effect ply forms alternating unclosed hook along both sides of yarn while the binder holds it in place.

Spike or Knarl

5. Boucle Yarns: They have light loops projection rom the y of the yam at fairly continuous intervals. The effect yam holds the I s in pas ti n. Boucle are soft with a rough surface. They can be utilised for km or wea .g.

6. Loop or Curl Yarn: It has closed loops at continuous in ervals al n the yarns. The effect of yam which forms loops or curls is made of either a single or a ply of two or more singles. It is comparatively fine has a s amount of twist. The loops are held in place by one or single yarns that work as binders.

7. Chenile Yarns: They are used for the manufacture enil The yam corresponds to a heavy caterpillar.

8. Metallic Yarn: They are mainly made by surr u ing a of aluminium or between two layers of simple yarns or palye the film is then cut.

III. Bulk Yarn

They are made from filament fibres. These are smoot d sli to touch. They lack the warmth, bulk and comfort. These yare from staple fibres.

Texturing of man-made filament fibre can produce :

— Bulkier yarn

— More absorbent yarn

— Yam with increased warmth

— Softer Yarn

— Yarns that do not pile

— Yarn with increased comfort

— Yarns that can stretch

Textured yarns are commonly divided as—

1. Stretch Yarns: Stretch yarns have high intensibility and better recovery .Bulk is moderate and is utilised in stretch fabrics.

2. Modified stretch yarns: Modified stretch yarns are subjected to hot treatment and are more stable than stretch yarns.

3. Bulked Yarns: They have moderate stretch. Stretch are utilised in sweaters, warm hosiery carpets and upholstery .Bulk texturing can be utilised with any type of filament or spun yam. Bulk texturing is done by:

(A) Gear crimping-Saw-teeth crimp.

(B) Stuffer box process-Saw teeth crimp.

(C) Air-jet process-loops in the fibres.

A. Gear Crimp : The filament yarns are passed between the teeth of two heated gears that provide the yam the shape of the gear teeth. Saw-teeth crimp with thermo- Gears plastic fibres the crimp is permanent.

B. Stuffer Box Process : This method produces a saw tooth crimp of significant bulk. Straight filament yarns are literally stuffed into one end of a heated box.

They are then taken out from the other end in crimped form. Bulked single yarns are generally piled to hold the Gear Crimp filaments together crimp. Thus the clear volume increases. It is fast method an is widely used. This crimped yarn is utilised for carpets, upholstery, hosiery and knitting.

C. Air-jet Process : This method produces loops in the yarn. 'Yarns are fed through an air-jet roller that moves more grodually than the stuffer box process.

The air-jet forms loops in the yarn. These loops ehance the bulk of the yarn. Any type of fibre can be utilised. Air-jet yarns maintain their size and bulk under tension since the straight section of the fibres bear the strain.

Felting

Wool has been utilised for centuries in making felt. Fur fibres may be utilised in felts for making hats, covers, toys, mats and diverse other purposes. Man-made fibres like rayon can be mixed with wool to make felt. The surface of the wool fibres is covered with short scales. The scaly structure causes wool fibres to cling closely together, as the scales from one fibre interlace with the scales of another fibre.

The material crimp of wool helps in felting. When masses of wool fibres are placed together, the crimped fibres become entangled and friction increases. this entangling. If the wool is subjected to heat and moisture, the scales open wider and interlock with more scales from the neighbouring fibres. The moisture causes the fibres to shrink together so that the mass is more tightly held together. When pressure is given the mass is flattened yielding a web of tightly joined fibres or felt. Thus heat moisture and pressure are needed to make felt from wool.

Production of felt

Two webs of carded fibre are laid across each other with the fibres of one web at right angles to the fibres of other web. Through the mass of fibre steam is forced. Then a heavy heated plate or roller is lowered on the fibre. The roller is moved about to produce friction. The moisture heat and friction effectively interlock the fibres.

After this the fabric is passed through soap solution or acid that causes the fibre to shrink more. The fabric now goes to the machine in which the fabric is subjected to further agitation, pounding and shrinkage.

After felting is complete the fabric is dyed or provided any flnish. Felt can be smoothly cut and because it has no yarns, it does not fray at the edges. Felt can be moulded into shape and is therefore used in making hat$. Felt is warm and water cannot infitrate it. It is a weak fabric, tears under pressure and is subjected to piling. It is stiff and therefore has limited uses.

Knitting

Knitting is a process of making cloth by forming a single yarn or a set of yarns into a series of interlocking loops.

Instead of two sets of yarns crossing each other as in weaving a yarn in knitting is looped through itself through many columns of stitches which are inttrelated but in horizontal methods ofknitting and verticle direction. There are two cardinal forms of knitting-

(A) Weft knitting (B) Warp knittng

A. *Weft knitting*: In this method a single yarn is fed into a number of needles in the horizontal direction. The cloth is made by building. 100 on top of each other.

On either a circular or flat by the machines weft knitting can,be done. Weft knitting is also made by hand knitting.

B. *Warp Knitting* : In warp knitting a large number of parallel yarns are mounted on the knitting machine. The yarn runs along the leagth of the fabric. These yarns form vehicle loops. These loops are made small easily by pulling.

Braiding

From yarns braided fabrics are made. They are made by interlacing (plaiting) three or more yarns that originate from a single lo ation and lie parallel before the interlacing occurs. The yarns in er-cross from one side to the other, resulting in a column of horizontal Vs verticle

To form wide fabrics or large articles such as rugs narrow braids can be joined together. They are made in flat rectangular or circular formations. Circular braids are spun in shadows and insulation for wires. Flat braids are utilised for trimming. Braided fabrics have great stretch in length and width. Cut strips of fabrics, leaves, leather, straw can also be braided.

Netting

Nets are the open mesh fabrics. They are prepared of silk, cotton, nylon polyester rayon or other man-made fabrics. They are made by looping and knotting a regular strand of yarn into an open mesh. Previously they were made by hand but now they are made by machines. Nets can be made on terricot and several other yarns.

Bonding

It includes the yarning together of two or more fabrics by an adhesive flame foam process. The purpose of bonding is to give greater stability and body to the face fabric or to generate a self-lined fabric. The two layers are joined by an adhesive, a binding agent or heat. The foam flame process of bonding is performed by melting a thin layer of poly urethane foam by passing it over a flame and then the two layers of fabric are sandwiched around the foam which dries and makes the bond between the two layers. Bonding produces warmth without weight. It makes it possible the use of right weighing fabrics for out wear. Bonded fabrics do not hold sharp creases.

Weaving

It is the interlacing of two sets of yarn which inter-lace at right angles to .each other. The length-wise threads are known as warps, individually they are called ends. The cross-wise threads are known as filling or weft; individually they are known as picks. Weaving is done on a loom which makes it possible to interlace the warp and filling threads according to a pre-arranged manner. The length-wise edges of the fabric are called selvages. .

Major Parts of a Loom

1. Warp beam: A roll or cylinder it is around which the warp threads or ends are wound in a uniform and parallel arrangement.

The unwinding of the yarn from the warp beam is called the "let off motion".

2. Whip roll : This is guide roller which directs the warp threads on their way to the lease rods and heddles.

3. Lease rods: An another guiding device it is for the warp yarns. These are two sorts of wooden or glass rods set between the whip rolls and the heddles.

Alternating warp threads can be kept separate by passing over and under these rods.

4. Heddles : The heddles control the warp threads like puppets. They are small steel wires set vertically into the harness frame. Each heddle bar controls one warp threads which is threaded through an eye in the centre of the heddle.

5. Harness: A wooden frame it is which they produce the shed or space between the warp threads, so that the filling yam can be passed through. This is known as shedding motion.

6. Bobbin and Shuttle: The filling thread is wound on a bobbin which sets into a shuttle or bobbin container. As the shuttle passes back and forth through the warp shed, it realeases thread from the bobbin and so forms the filling of cloth.

7. Reed: This is inevitately a combination made up of steel wire rods set vertically in a frame. 11te spaces between the wires are known as splits, or .dents and are kept even and paralled. This is the first function of the reed.

Its second function is to feed the filling thread into position. To do this it has to move in a back and forth motion.

8. Breast Beam: The bar this is, at the front of the loom over which the cloth passes on its way to the cloth roller.

9. Temples: They divide at the edges of the cloth which supports to maintain fixed dimension in width.

10. Cloth Roller: It is the roller on which the newly made cloth is rolled on the cylinders.

Characteristics of Woven Fabrics

I. Count of cloth or thread count : The yarns of warp and yarns of filling is not incessantly of the same diameter. Also those utilised in one direction may be closer together than those of the other direction. Often there are more warp yarns than filling yarns to the inch since the strain on a fabric falls on the warp.

Some cloths like gingham are close woven while those like voile are loose woven. A closely woven fabric keeps its shape shrinks-less, slips-less at the seam and is more durable than a loosely woven cloth. The closeness or looseness of the weave is measured by the count of the cloth. This is conditioned by the number of pick and ends to the square inch be changed by shrinkage during drying and finishing.

With the warp number thread count is written first, e.g. if there are 60 warps and 50 fillings to the inch then the count is 60 x 50. Thread count indicates the quality of a fibric. There are high count and low count clothes. The count is made with a thread counts. A small pocket magnifying glass, called a pick glass or linen tester is used for counting—

(a) Many warp yarns and many filling yarns are removed from the cloth.

(b) If the fabric is light in colour a cut of black material is put under it or vice versa.

(c) The tester is put on the raveled edge.

(d) Since the opening in the tester is 1/4" square, the yarns are counted in this (first number of warps then of filling).

(e) Then number of yarns that run every way is multiplied by four to receive count per inch.

(f) A pin sometimes support to separate the yarns.

I. Balance of cloth :

I. The ratio of warp yarns to filling yarns is known as the balance of a cloth.

2. If the number of warps and number of fillings to the inch are nearly the same (not more than 10 yarns difference) a cloth is said to have a good balance for instance if the count is 60 x 50, it would be a well. If balanced cloth. If the cloth has poor balance than the warp yarns have a tendency to slip over the filling yarns. Shirts, pillow slips and towels which have various washings must have a good balance.

II. Selvage :

A selvage is the self-edge of a fabric formed by filling yarn when it turns to go back across the fabric. They are made in many ways :

(a) Plain selvages: These are corresponding to the rest of the fabric. They do not shrink and can be used for seam edges in garments construction.

(b) Tape selvages: They are prepared of larger and for ply yarns to give strength. They are wider than the plain selvages.

(c) Split selvages :They are used when narrow items, such as towels are made by weaving two or more side by side and cutting them apart after weaving. The cut edges are finished by a machine, chain-stitch or hem.

(d) Lased selvages: They are the heat sealed edges of ribbon generally utilised for synthetic fibres.

The Three Basic Weaves

I. Plain Weave

The simplest of the three basic weaves. This is Most woven goods are prepared in plain weave. It is the strongest of all weaves since its threads are interlaced the highest. The plain weave is the simplest form of inter-lacing the warp and fillings yarns. To make this weave, the odd numbered ends are raised over the first pick and the even numbered ends are raised over the second pick. Plain weave fabrics have no right or wrong side till they are not printed or given a finish. Plain weave fabrics tend to wrinkle readily and are less absorbent than other weaves. By using various fibres, novelty or bulk special effect can be provided to the plain weave fabric.

Variations of Plain Weave :

A. Rib Weave: To make a plain weave fabric more attractive a ribbed or corded effect is produced by using warp much heavier than fillings as in the former method is the most common.

A striped effect is produced by alternation of fme and heavy warps at continuous intervals as in striped or corded madras shirts.

Fine and heavy fillings may be alternated to produce a crossbar effect, e.g. crossband weave Iis tissue girghans.

B. Basket Weave: The basketvariaon of the plain weave is attractive and guides design to the plain weave. It is not as durable as the plain weave and has low strength. One or more filling yarns are passed alternately over and under two or more warp yarns.

If two warp yarns are pass over and under one fining, it is known as a 2x I basket. Other fabrics of the basket weave can be found in 2x2, 2x4 and 3x2 basket but the 2xl basket is most common. It is loose, slippage of yarns takes place. It may shrink readily on washing, e.g. of basket weave are monks cloth, oxford cloth etc.

II. Twill Weave

Twill weave is the second basic weave that can be made on the simple loom. It consists of diagonal bars which gives a sturdy look to the fabric. These diagonals, known as wales, may run from upper left to lower right {left hand twill) or from upper right to lower left (right hand twill) or both ways in the same cloth as in herring Bone. The warp yarn floats over two filling yarns and under one filling in the 2/1 twill.

When the warp yarn floats over two filling yarns and under two fillings, thus 2/2 twill is formed.

The ridges or wales are formed towards right or left since warp and weft are in a incessanl manner. Twill weaves are closer in texture, heavier and sturdier than plain we ayes and can be produced in several fancy designs. The number of pick yarns needed to couple a twill pattern conditions the number of harnesses loaded on the loom. There must be at least three; and some twill patterns require as many as fifteen. More the harness to be added, costlier it is.

Twill Weave: Right hand side twill/Left hand side twill. Instances of fabrics made by the twill weave comprise denim, drill, zean, gabaradine, flannel etc.

Variation of the Twill weave Herring Bone.

In this design the twill is reversed at frequent intervals to found a series of inverted wales. III. Satin Weave

The Satin weave is the third basic weave that can be made on the simple loom. Satin weave is characterised by long, floating warp yarns that produce a high lustre on one side of the fabric. Warp yarns of low twist float or pass over four or more filling yarns in a continuous pattern.

The filling yarns that appear on the surface are spaced so as not to exhibit a diagonal pattern in the low twist, and the floating of the warp yarns. Generally filament fibres give a high degree of light reflection, which adds to the lustre of the fabric. The long floats are susceptible to abrasions and sagging in wear, and therefore satin weave is not as durable as twill or plain weave, If the fabric is made with a high number of yarns per inch (high count), it will be relatively durable. Instance of fabrics with satin weave are satin. nylon satin, slipper satin, and crepe backed satin.

Sateen: In this weave the floating yarns float and are explibit on the surface. Spun yarns with low twist are utilised for this weave and hence the fabric has the tendency of getting worked up. This makes the fabric fragile and not fit for the daily use.

Bobtex Produces a yarn by pressing staple fibres of any length into a molten polymer stream.

III: Fancy Weaves

The Fancy Weave is the fourth basic weave that can be made on the simple loom. Some of the variations of fancy weave are-

(A) Pile Weave.
(B) Ornamental Weave.
(C) DobbyWeave.
(D) Leno Weave.
(E) Jacquard weave.

These weaves are produced to make more detail patterns and provide them rich and diverse effects. The durability relies upon the length of the floats and the yarn utilised.

A. Pile weave: Woven pile fabrics are of the three-dimensional structure. It is made by weaving an extra set of warps or filling yarns into the ground yarn to make loose or cut ends on the surface. The right side of the fabric comprises of soft clipped yarns known as pile. The wrong side of the fabric is smooth, with no pile and the weave (plain or twill) showing, distinctly.

Cloths with pile are velvets, corduroy and velveteens.

There are five methods of making a pile-

(c) The Wire Method.

(b) The Terry Weave Method.

(c) The Filling Pile Method.

(d) The Double Weave Method.

(e) The Rug Method.

(a) The Wire Method: Velvets, plushes and rugs are made with additional warp to form the pile. One set of warp yarn interlaces with the filling to form the plain or twill weave ground of the fabric. The other set of warps forms the pile. When a row of pile is made, the warp yarn", to form the pile is raised by the harness to form the shed.

Then a wire is included through the shed just like the filling. The size of the wire used relies on the size of the pile to be made. When the set of warps to form the pile is lowered. it loops over the wire and is held in place the next filling. The wire is then repl and as this is done a small sharp knife attached to the end of the wire cuts the pile warp loops. The ground is then woven for a particular number of picks. then the wire is again included to form the pile. If the pile has not been cut equally by the , wires, the fabric is sheared again with a device like a lawn mover.

(b) The Terry Weave Method: This is cheap method. The warps that form the pile have their tension released at intervals and are then pushed forward. After this the tension is restored. and the beating of the filling causes these warps to look in loops.

The pile is generally on both sjdes of the fabric. Turkish towels are woven in this manner. Loops are uncut in Turkish towels to enhance absorption.

(c) The Filling Pile Method: To make corduroy velvet chenille and some plushes, extra filling are floated over four or five warps. The floats are cut after weaving and the cut ends are brushed up to form the pile. Corduroy floats are interlaced in such a manner that rows are formed and when the pile is cut it produces a ribbed effect. The circular knives revolved round and cut the floats to form a pile.

Valley left under floats after cutting, becomes lines between ribs.

(d) The Double Weave Method: Two cloths are woven face to face at the same time-. Two sets of warps and two sets of fillings are utilised and an extra-set of warps binds the two cloths together. The plain, twill or satin weave are utilised as a ground.

The Double Weave Method

After the cloth is woven a knife in the loom cuts the binding yarns making two separate fabrics with sheared pile surfaces. Generally transparent velvets are woven by this method.

(e) The Rug Method: Hooking, tufting, tagging, chemical and buried pile are way of making pile for rugs and carpets. Sometimes the back of the fabric is bonded with the layer of synthetic resin in rugs.

B. The Jacquard Weave : Detail designs are woven on intricately constructed looms called Jacquard looms. In the piece of harness, a number of oblong, punched cards are utilised which control the raising of warps. The design is first worked on the graph paper in points.

It takes weeks to make the loom for intricate patterns. Jacquard weave fabrics are usually purchased for their beauty and not for their durability . Since the design unit repeately has long floats, the fabrics snag easily, dearth dimensional stability in laundering and dry-cleaning and the stains are difficult; to remove.

Examples of Jacquard weave are brocade, damask, tapestry, etc. The weave is actually a combination weave, Two or more of basic weaves are combined in the same cloth. For instance, in damask, the design may be a sateen weave and the back-ground may be a satin weave.

C. Dobby Weave : Small designs can be woven inexpensively by placing a dobby attachment on a plain harness loom.

Simple small geometrical designs or figures are repeated generally throughout the fabric (e.g: honey comb birds eye). Used for diapered. The dobby attachment cannot control more than 25 harnesses hence only limited number of designs are available.

D. Leno Weave : It is made by particular attachment to the loom. The warp yarns do not lie parallel to each other instead, yarn of each pair is crossed over the other to form a figure of' 8' before the filling yarn is included. The filling yarns in a leno weave are held in the place strongly and cannot shift in the warp direction. But the warp yarn can shift in the filling direction.

The construction may be varied by the number of filling yarns inserted. Fabrics made of leno weave are open textured and may vary from sheer to heavy. The leno weave Ipay be combined with other basic weaves to make good patterns. Instances of leno weave are mosquito nets, mosquito sette etc.

E. Ornamental Weave : Patterns similar to embroidery can be woven into cloth at the time ground work is woven. The difference between these patterns and others—is that they can be pulled out without destroying the ground work. The shrivel and lappet design shall only be discussed.

(a) Lappet Weave: Lappet weave is a kind of weaving in which extra warp threads create design that corresponds embroidery .The pattern seems only on the right side. The designs are made of one continuous yarn and are not clipped. It is a strong weave but is expensive. It may be used on all types of cloths for many uses.

10

Clothing Care and Sewing Techniques

Threading a Needle

A. Self-Threading Needles

Self-threading needles have two holes, one of which is open. The thread is placed across the top and pushed down through a slit into the second hole. There are several hand positions that can be used to thread the needle while holding it in your hand. The recommended method is to place a pin cushion on the table with the needle stuck straight into it. This allows two free hands to hold onto the thread. Wrap the thread around the two index fingers and use the thumbs to locate the position of the needle in the pin cushion, and then push the thread down onto the needle. A disadvantage of self-threading needles is that the thread pulls out of the slit in the needle very easily.

B. Wire Needle Threaders

Wire needle threaders can be purchased in packages of three

for about $1. To use this, push the triangular shaped wire through the eye of the needle, put the thread through the triangle, and pull the wire back through the needle with the thread. These may be difficult to use for some users because the wire has to be pushed through the eye of the needle first and the wire may bend or a user may not find the hole in the needle. F the user to wrap the thread around a straight pin and stick the straight pin through the wire loop. This gives the user something stiff to stick through the wire triangle instead of the limp thread.

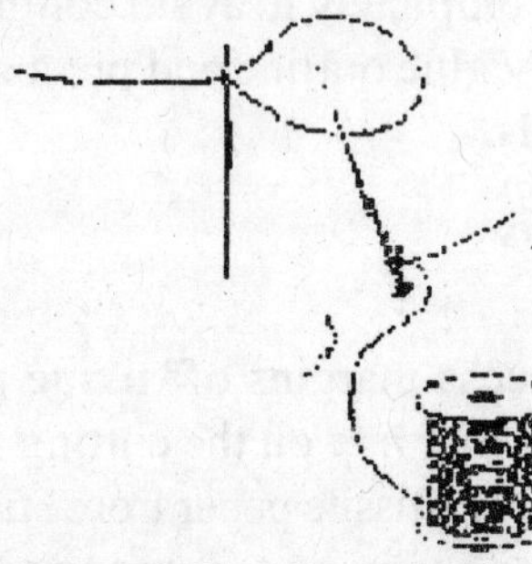

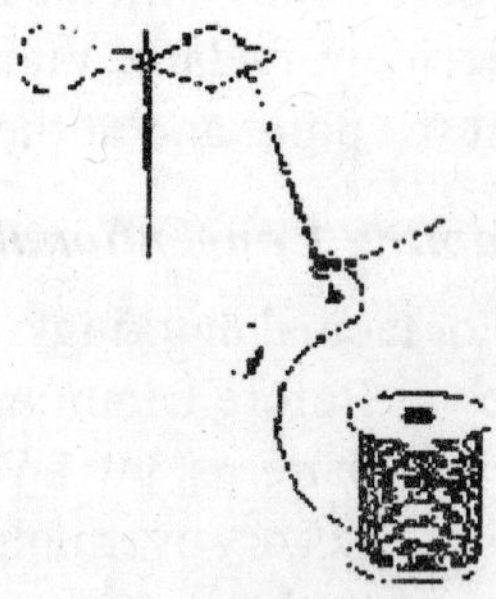

Dental Floss Threader **Wire Needle Threader**

C. Dental Floss Threaders

Dental floss threaders can be purchased in packages of 20 from the drugstore for about $2 - $3. user prefer these threaders over the wire needle threaders because they are stiff enough to easily thread into the eye of the needle and the loop is flexible, unbreakable, and large enough to feed the thread through. The loop is pulled through the needle from front to back.

Patterns and Fabrics

A. How to Make Your Own Patterns

For making patterns out of heavy brown paper or butcher paper, using conventional tissue paper patterns as guides or creating your own:

- Outline the edges with a 1/4 inch line of Elmer's glue.

- Mark the straight grain line arrows with a strip of 1/2 inch masking tape or brightly coloured labeler tape.
- Put a strip of masking tape folded over the paper edge to identify the lay on the fold.
- The crosswise grain line arrow can be marked with a strip of masking tape with staples at each end.
- The bias grain line can be a strip of masking tape with staples lengthwise along one side of the tape.
- Use staples to identify the location of notches. For beginners, I sometimes omit the notches completely to avoid confusion.
- To identify darts, make a heavy glue outline and put a staple at the point and at the two ends.

B. Adapting Conventional Patterns

1. For tactual markings:
   - o Before trimming any excess margins off tissue paper patterns, put an Elmer's Glue line on the cutting lines. Let dry overnight. To keep the tissue paper from sticking to the counters, puts scotch tape on the wrong side of the pattern piece along the cutting lines.
   - o Mark these patterns with masking tape as described above.
   - o The student should be able to tear off the excess margins outside the glue lines before pinning the pattern on the fabric. Otherwise, cut off the excess margins outside the glue lines.
2. For visual markings: Use a colour broad tipped felt pen and retrace all the cutting lines and special markings, such as darts, dots, and pocket placement lines, etc. It is easier to do this before the excess margins have been cut off of the tissue paper pattern.

C. Buying Fabric

Students should be encouraged to learn to differentiate textures and weaves of various types of fabrics (e.g., cottons, double knits, wool). A trip to a fabric store should include opportunities to tactually explore fabric. Describe the characteristics of different fabrics verbally to the students. Help them pick a pattern, or practice soliciting help of a sighted person, and practice buying the fabric and all the sewing notions needed, and paying for the purchase.

D. Transferring Patterns Fabric

Make all necessary adaptations (e.g., glue lines or coloured felt-tip lines). From this point on, the user should be able to lay out the pattern (using the masking tape grain line arrows as guides), pin the pattern to the fabric, and cut out the pieces. For beginners, recommended simple projects, such as an apron, tote bag, poncho, skirt with elastic in the waistline, pants with drawstring or elastic in the waistline or a "torn project" (where the user tears out the pieces rather than cutting out the pieces). It is recommend that the fabric chosen be a woven polyester and cotton fabric.

Only special markings such as darts, dots, pocket placement lines, or notches need to be marked on the fabric. Darts can be identified by placing a small safety pin at the point and at each end, where the staples were on the adapted paper pattern. When the darts are ready to be sewn, the safety pins are replaced with straight pins.

Machine Sewing

A. Pinhead Guide

The use of a pinhead guide helps the blind student sew straight. Place a row of straight pins horizontally onto the sticky side of a piece of masking tape. Let the heads of the pins extend over one side. Place another piece of masking tape with the non-sticky side directly over the pins. Stick the pinhead guide to the metal slide plate on the sewing machine.

The placement of the pinhead guide may vary depending on the task (regular seam at 5/8 inch, stay stitching at 1/2 inch, topstitching at 1/4 inch). A notched metal seam gauge, similar to a 6-inch ruler, but with indented notches at each half-inch mark, can be used to aid the placement of the pinhead guide. Most sewing machines come with an etched line on the metal slide plate that marks the 5/8 inch regular seam line. Most students can feel this line and place the pinhead guide on it. When beginning to sew, the fabric is lined up with the first pin on the pinhead guide (which is even with the machine needle).

B. Metal Seam Guide

Most sewing machines have a screw hole to the right of the needle for a metal seam guide that functions like the pinhead guide. These metal seam guides are also available in a magnetic form. The screw on type because the magnetic ones tend to move out of position. Some user use a combination of the screw on type seam guide and magnetic guide (or pinhead guide) in order to make a longer edge for the fabric to move against.

C. Needle Finger Guard

This is a small 3-sided metal bar attached to the sewing machine on the same bar as the presser foot near the needle. It is a safety device to warn the user that the needle is close by. The finger guard is pulled down in front of the needle when sewing and pushed up to the left of the needle when threading. Most new sewing machines have a finger guard.

D. Darts

If the darts were marked with safety pins, fold the dart in half by making the safety pins even. Replace the safety pins with straight pins and put a straight line of pins from the wide end to the point by using the seam gauge or ruler as a straight edge. Place the pinhead guide directly in front and in line with the needle and presser foot. Start at the wide end of the dart (at the fabric edge) and begin stitching, holding the pins in the fabric flat against the pinhead guide and remove the pins one by one to the point of the dart.

Hems

Mark the length desired with a pin. Measure the fabric to be turned up with the notched seam gauge and pin up the entire hem. Press the pinned up hem.

A. Sewing Hems with the Sewing Machine

Whenever possible, use the sewing machine to sew in hems.

Use the metal seam guide or pinhead guide, placing it as far from the needle as the depth of the hem, and stitch.

B. Ironing Hems with Stitch Witchery

This is an iron-on adhesive, mesh-like material available in strips or by the yard. The strips are easiest to use for hems. After pressing the hem up, remove the straight pins and place the Stitch Witchery between the fabric and iron.

C. Sewing Hems by Hand

Thread a needle using a double thread. A single thread comes off the needle too frequently. Hold the hem in one hand with the thumb on the pinned hem edge. With the other hand, put the needle through the fabric until the point just touches the index finger and then push the needle back up through the fabric. Position the thumb so that the first stitch is on one side of the thumb. Now take the second stitch on the other side of the thumb. Continue around the hem, using the thumb as a guide for the size of the stitch to make.

Ironing Clothes

A. Safety Techniques

First, you have to learn how to turn the iron on and off, the positions of basic temperature settings, how to add water to steam irons, how to set the iron down when it is hot, and which parts of the iron get hot. With a cold iron you have know how to hold the iron correctly, to put the iron down by keeping the forearm straight out with the elbow next to the body. To find the iron again after putting it down, trail up the ironing board on the side closest to the body or find the cord and go up the cord to the handle.

B. How to Iron a Shirt

- Iron the collar first by putting the collar wrong side up on the ironing board with the seam on the outer edge of the ironing board (side away from body). Hold it in place with your hip and pin the corners of the collar down with straight pins, then iron. Remove the pins.

• Iron the shoulders and, if applicable, yoke by inserting the tip of the ironing board. Hold at the collar and at the bottom of the yoke or pin the yoke down (with the collar closest to your body). Remove the pins.

• Iron the sleeves by finding the underarm seam and folding along that seam. Pin the sleeve to the ironing board and iron. Push the iron sideways toward the collar to feel when the side of the iron reaches the armhole seam (and avoid ironing the collar again). Remove the pins.

• Iron the body of the shirt, starting with the button side of the front. Pin the tail of the shirt down and hold onto the collar as you iron. When finished ironing a section, unpin it, slide the far edge of the shirt at top and bottom toward the edge closest to your body to position an un-ironed section. Pin this un-ironed section down and iron. Continue in this fashion around the shirt. When ironing the body of the shirt near the armholes, slide the shirt so that the end of the ironing board is sticking inside the top of the sleeve. This will keep the shirt flat on the ironing board.

Folding Clothes

A. Folding Shirts

First button every other button and the cuffs, if applicable. To find the top of the shirt, put your hand inside the top of each sleeve. Shake out the shirt holding onto the top of it. Lay shirt face down on a table with the collar to the left. Pull the sleeves out to the sides increasing the body as much as possible. Bring the body of the shirt close to the edge of the table. Put the index finger side of the left hand next to the collar (right side of shirt), fingers pointing to the shirt tail, and fold over right side of shirt. Fold the right sleeve lining it up lengthwise with the shirt. Repeat for left side of shirt. Bring the tail end of the shirt up to the collar to fold in half lengthwise, or fold in thirds by bringing the tail end up one third and then fold again in half.

B. Folding Creased Pants

By holding the bottom of each pants leg put the seams together, making sure the inside seams are touching. Hold the bottom of the legs and put under your chin. Bring the waist of the pants up and fold in half, or hold on to each end and bring hands together.

Hanging Clothes

A. Hanging Shirts and Dresses

Put the hand inside each sleeve or armhole and slide the hanger in. Button any top button to prevent the garment from sliding off. Line up either the tag on the shirt with the hanger hook or the shoulder seams with the hanger arms.

B. Hanging Pants

Crease the pants the same way as when folding pants. Lay the pants flat on a table with the pant legs out in front of you. Slide the hanger under the pant legs, almost to the crotch, and raise up off of the table.

Laundry

A. Washing and Drying

Most washers and dryers have various types of settings. If possible, mark the settings for regular and permanent press with glue dots, puffy paint, or Hi Marks. Do not overload the washer or the dryer! Teach how to sort clothes according to colour and type of laundry and how to measure detergent. Remember to put the detergent and bleach into the washer with the water **before** putting the clothes into the washer (it is a good idea to swish the water a few times to dissolve the detergent). To avoid ironing, take polyester clothing out of the dryer before completely dry and hang up, or as soon as the dryer stops, hang up the clothes quickly.

It is not recommended for blind students to use bleach in their laundry. If bleach is to be used, then use it only on white cottons (make sure the user knows about different fabrics). An alternative would be a detergent with bleach safe for colours.

B. Stain Removal

Stains or spots on clothing must be found or identified by a sighted person. The stain should be marked with a safety pin, or if the spot is large surround the spot with safety pins. Use a prewash or stain treatment and let it soak for a few minutes. Then wash the garment with the other laundry in the washing machine.

Sewing Techniques

Uses of Bias Binding **:** In this lesson we are going to study all about bias bindings, how to make them and how to apply them to the pretty clothes we are planning to make very soon. Study the lesson carefully—it contains some very valuable information for you.

Cutting guage on your sewing machine **:** Perhaps you do not know that in your set of attachments that came with your sewing machine there is a little attachment that saves you lots of time and trouble. We refer to the cutting gauge which enables you to make all the bias strips you want right at home. Before anything else, we want you to know just how to use this valuable gauge.

1. Remove your cutting gauge from your box of attachments. Do not become alarmed if you do not have a box of attachments—if you lost, mislaid or never received it. Simply write to the sewing machine company, ordering a box and also a booklet so that you can easily identify them.
2. You will find the gauge marked to cut different widths of binding. There are three letters on the gauge, F, B and C.
3. When set at F the strip cut is wide enough for facings; when set at B it is wide enough for bindings; when set at C the strip is just-right for cording. Read this over again and try to remember it.
4. Cut and join Bias pieces After these pieces are cut and joined adjust the gauge to your scissors and cut these strips into narrow strips which will just fit the binder. If the gauge is set at letter B the binding will be the correct width.
5. Join selvedge edge to selvedge edge, crosswise edge to crosswise edge, and you will have your material ready to use in the binder.
6. Join the pieces and press.

Note that illustration shows the right and wrong methods of joining the strips after you have cut them with the aid of your cutting gauge. When you have your bias material all ready, wind it on a card-board to keep it smooth, and you will have several yards of bias binding for use in this lesson.

The Binder

Another very valuable attachment in your box of sewing-machine attachments is the binder. Of course fine bindings can be made by hand, but the modern, time-saving method is to make them on the sewing-machine with the aid of the rapid little binder. With this tiny bit of mechanical steel you can make ten yards of binding in ten minutes. Think how long it would take you to do it by hand!

1. Remove the presser foot from your sewing machine and attach the binder attachment. Your instruction book will tell you how to attach it.
2. The binding you have just made will be found to fit exactly into the measuring gauge found on the binder.

   Fold the bias strip in the center lengthwise, and cut it to a point so that you can insert it into the binder easily.
4. The binder has a scroll through which the binding must pass until the point comes through the forward end and under the foot. As the binding passes through the scroll you will notice that the edges of it are turned with almost mathematical precision—more even and precise than you could possibly hope to turn it by hand.
5. Now take the material that is to be bound; the edge upon which the binding is to be applied should be held in the scroll of the binder.

6. Now proceed to sew, being sure that the stitching always comes close to the edge and that the binding is not stretched.

Perhaps this lesson on binding seems puzzling to you, as everything new seems puzzling at first glance. But really, it's the simplest thing in the world once you learn how to use your binder correctly. And how useful it becomes when you are ready to make pretty little clothes for the kiddies and neat tailored things for yourself!

So practice binding over and over again until you know it. Practice on cheap muslin with contrasting colours so that you can quickly discern your mistakes—if there are any. We want to help you in every way possible and we do not want you to hesitate for one minute if there is anything you do not quite understand and would like to have more fully described.

Binding Straight or Curved Edges

You will find binding straight edges a very simple matter. Curved edges are a trifle more difficult to bind neatly, and consequently require more practice. But first master straight-edge binding, and you will find that binding curved edges is not so very difficult after all!

1. To bind curved edges, be sure that the edge to be bound is well held within the scroll. It has a tendency to slip out; the result, as you can readily see, is an ill-bound edge that is not neat and that will succumb easily to the first bit of rough wear.
2. Feed the material into the binder only as fast as the machine feeds. Do not hold back either material or binding, but learn to feed both simultaneously.
3. When turning a curved edge, urge the material proper around gently with the fingers of the right hand, being sure that the edge being bound and the binding are in place and doing their work correctly.
4. After you have learned how to bind curved edges with your binder you will realize how many uses there are for the simple little binder. It will make quickly and easily, edges for your underwear and trimmings for aprons and children's clothing.

5. Another use for the binder is to make plackets for wash dresses and under-garments. It makes these plackets neater and stauncher than you could ever make them by hand. To make it, cut a slot five or six inches long in a piece of cotton material and fold in one straight continuous length. Then insert the edge of the material in the binder and bind until you reach the lower point of the slot. Stitch slowly at this point, turn the material you are binding to the other side of the placket, and continue as before.
6. Binding makes a most attractive trimming when used as a French fold. To apply, insert the binding material in the binder and place the garment under the binder itself. Stitch the fold in the position desired.
7. Instead of using bias binding, as made according to our directions in this lesson, you can use folded tape, military braid, ribbon, or any binding material of the proper width with your binding attachment. However, do not use any of these prepared bindings for practice work as it is unnecessarily extravagant; the "home made" binding will serve the purpose just as well.
8. As soon as you feel that you understand the binder, practice making scallops. We do not expect you to be able to make perfect scallops, or even neat scallops, at first, but if you practice whenever you have a chance, you will soon be turning edges like an expert and binding scallops as though it were the easiest thing in the world.

Read over the directions until you understand them, then practice and practice until you feel sure that you will never have any trouble with edges that have to be bound.

Binding and Piping Instructions

Binding is a way of neatening a raw edge using a separate length of fabric, usually in a contrasting colour.

Piping is a way to emphasize a seam or edge and can add strength to corners that might receive a lot of wear.

For both binding and piping strips of fabric cut on the bias are used as this makes them slightly stretchy.

Instructions on Cutting Bias Strips

- To cut bias strips fold the fabric diagonally so that a straight edge is parallel to the selvedge (the side of the fabric).
- Press this fold and use this crease as a guide to mark parallel lines for bias strips.
- To join bias strips place two lengths, right sides together, right angles to each other and stitch.
- Open out, press seam open and trim off the overlapping edges.

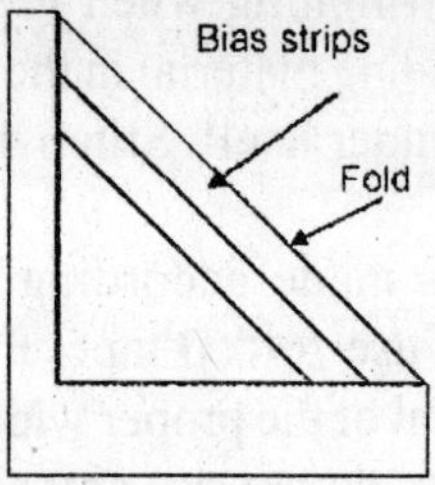

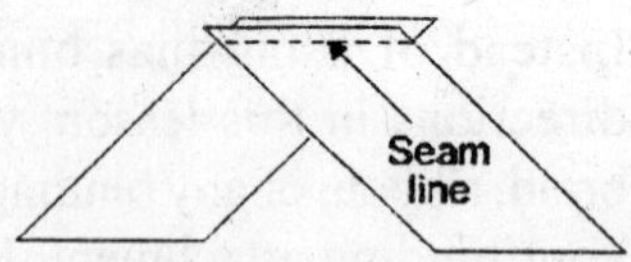

Instructions on Binding an Edge

- Cut bias strips 4 times the finished width of the binding required.
- Fold both raw edges to the centre on the back, down the length of the bias strips and press.
- Open up one side of the binding. Place this side of the binding onto the right side of the fabric with raw edges matching and pin.

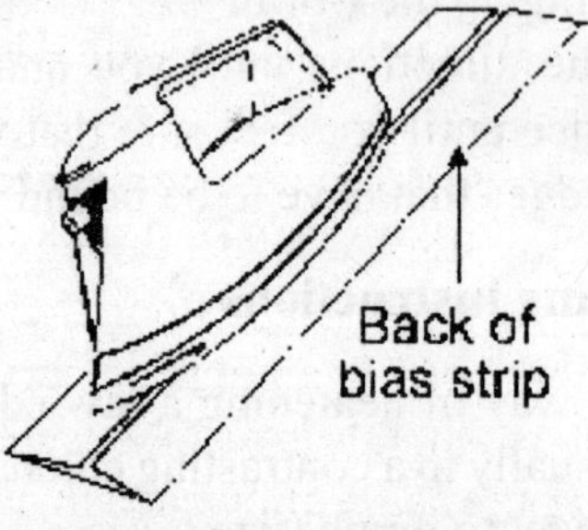

- Stitch along the fold line of the binding.
- Fold the binding over the raw edge to the wrong side of the fabric.

- Pin and slipstitch the fold of the binding to line of stitches on the wrong side.

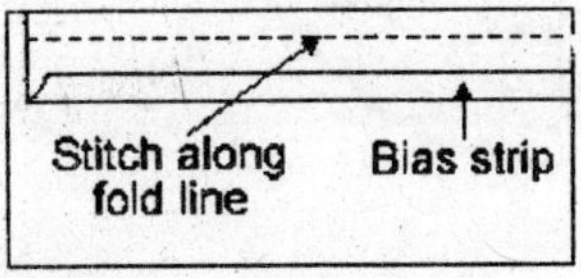

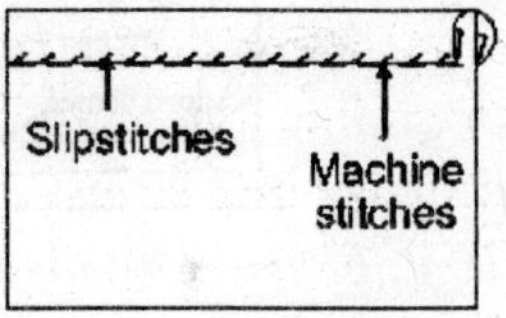

Instructions on Piping

Piping can be made by covering cord with strips of bias cut fabric. The cord comes in a range of thicknesses and is readily available from fabric and upholstery stores.

- Cut and join enough bias strips for the required length and wide enough to cover the cord plus the seam allowance.

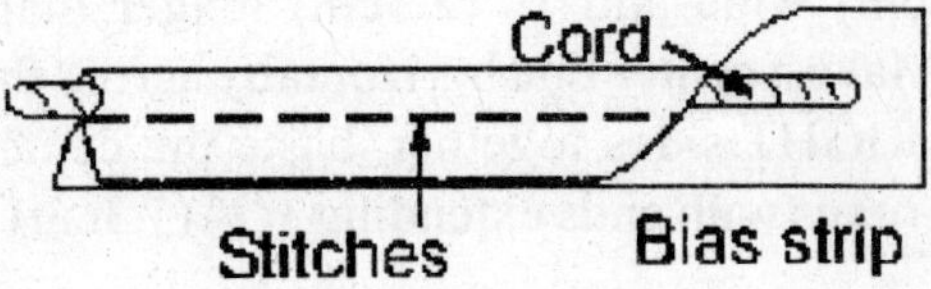

- Lay the cord in the centre, on the wrong side of the bias strip.
- Fold the bias strip over the cord matching the raw edges.
- Stitch the 2 sides of the bias strip together along the length. Use a zipper foot on the sewing machine to ensure that the stitches are close to the cord.
- To attach the piping cord to a straight edge, pin it to the right side of the main fabric matching the raw edges.
- To turn a corner, clip the seam allowance on the piping cord up to the line of stitches. Ease the piping cord around the corner and proceed pinning it along the next edge.
- To attach piping cord to an outside curve, place it onto the right side of the main fabric matching the raw edges. Pin it in place, clipping the seam allowance to ensure that the fabric lays flat.
- To attach piping cord to an inside curve, work in the same way as an outside curve allowing the seam allowance to overlap where it is clipped.

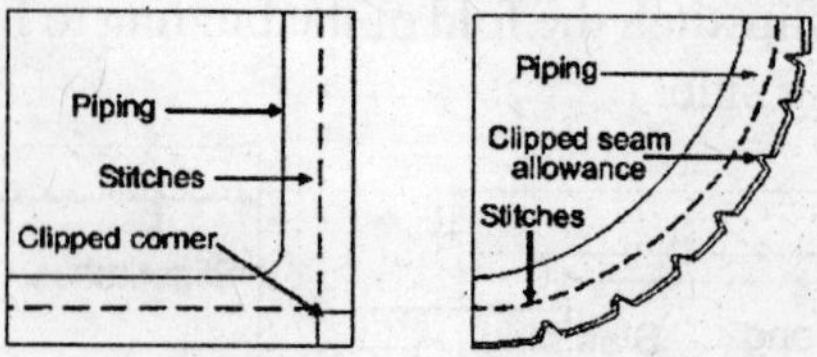

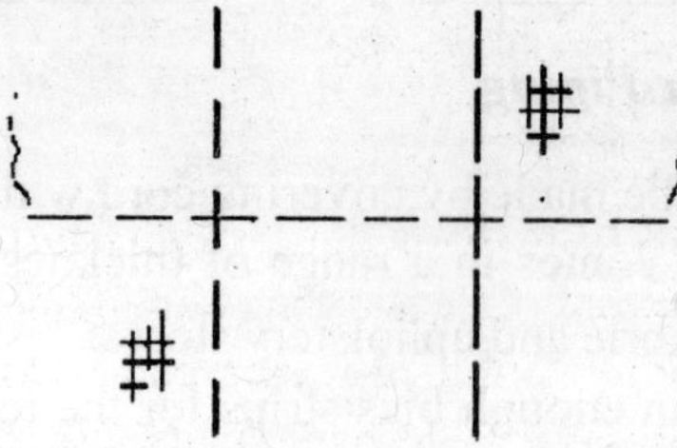

Step 1: Cut a self fabric patch for each buttonhole on straight grain, 2" (5cm) wide and 1" (2.5cm) longer than the finished buttonhole. Make a center line horizontally across the length of the patch. With RIGHT sides together, baste the center of the patch along ladder rungs with ends extending 1/2" (1.3cm) beyond ladder sides.

TIP: To create a bit of flair, consider cutting buttonhole patch from contrasting fabric.

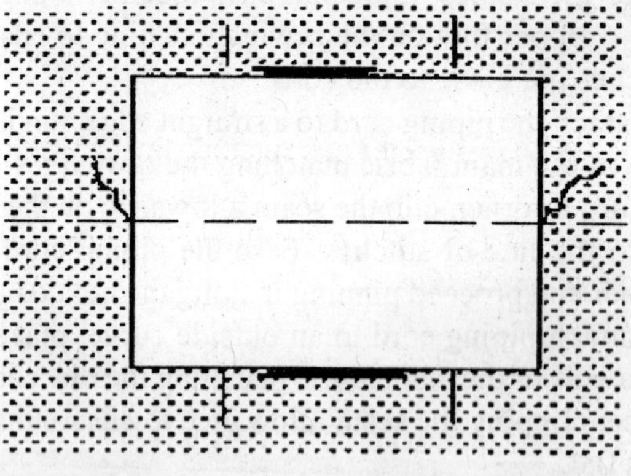

Step 2: On WRONG side, mark lines 1/8" (3mm) above and below buttonhole marking, using transfer pencil.

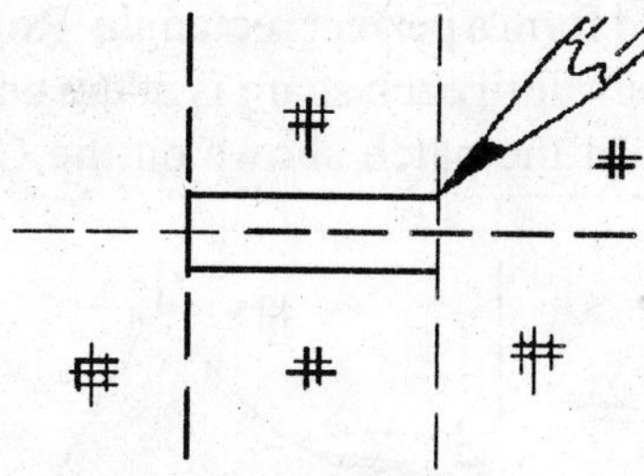

Step 3: Using a small machine-stitch, start sewing along one long side of the buttonhole, following pencil lines, being careful to end stitching exactly at thread-traced ends. Do not stitch across ends, back-stitch or pivot at corners. Bring thread ends to WRONG side and tie in a knot.

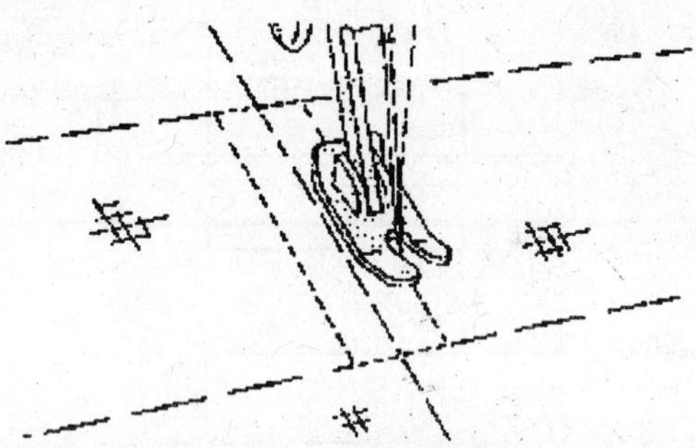

Step 4: Snip between the two stitching lines and clip diagonally to corners, as shown. Be careful not to clip through machine stitching. Add a dot of seam sealant (such as Fray CheckTM) to corners, and allow to dry.

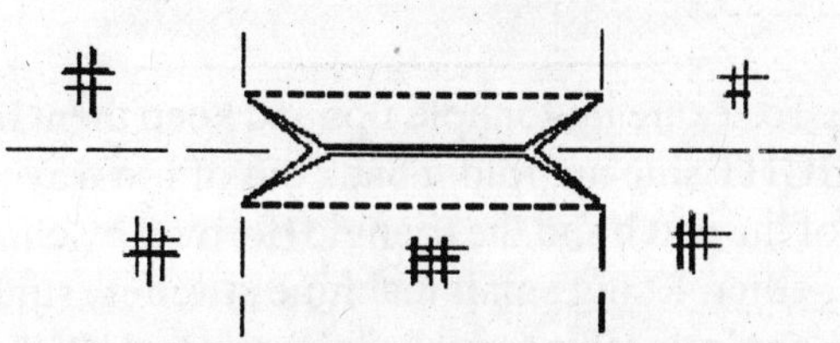

Step 5: Pull patch through opening to WRONG side of front. This opening should form a perfect rectangle. Roll edges of opening between your fingers until each seam is at the edge of the opening. Press so that none of the patch shows on the OUTSIDE.

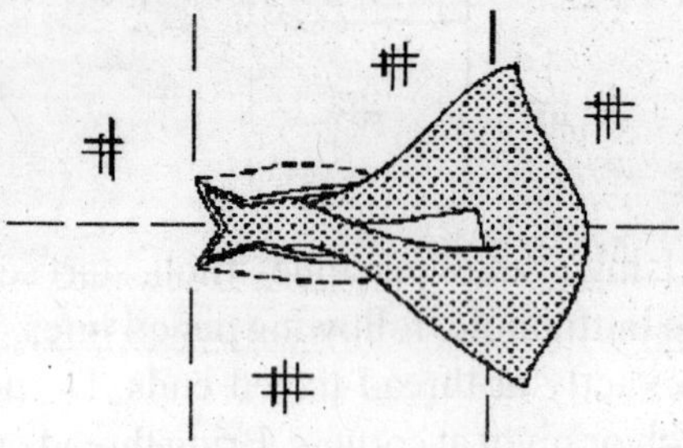

Step 6: To form buttonhole lips, fold each long side of the patch over the opening, so that the folds meet exactly at the center.

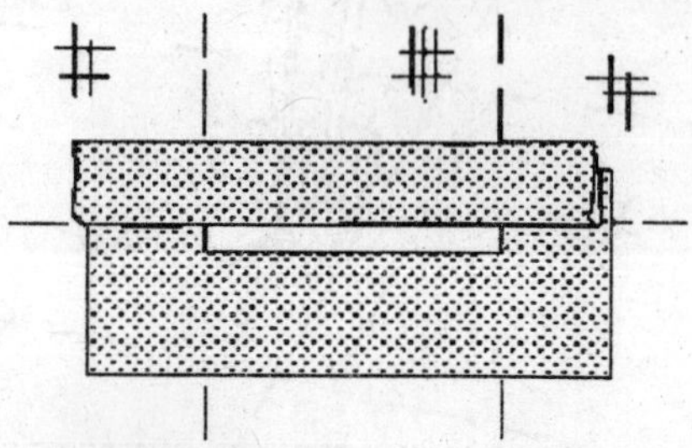

Step 7: On OUTSIDE, whip-stitch buttonhole lips together along fold lines and press in place. Whip-stitches should remain in place until garment is completed.

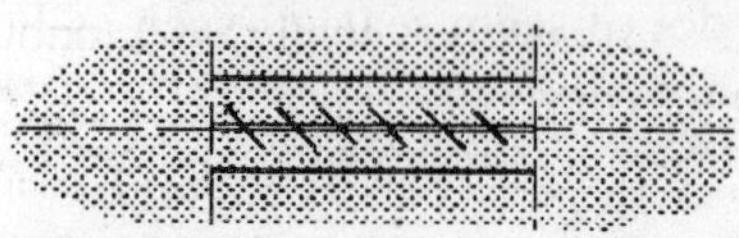

Step 8: To secure buttonhole lips and keep them from shifting, with front RIGHT side up, fold it back out of the way until you can sew the end of the patch and the triangle (formed by clipping corners in step 5) together. Using small machine stitches, stitch across the base of the triangle, catching patch. Trim end to 1/4" (6mm). Repeat on other end.

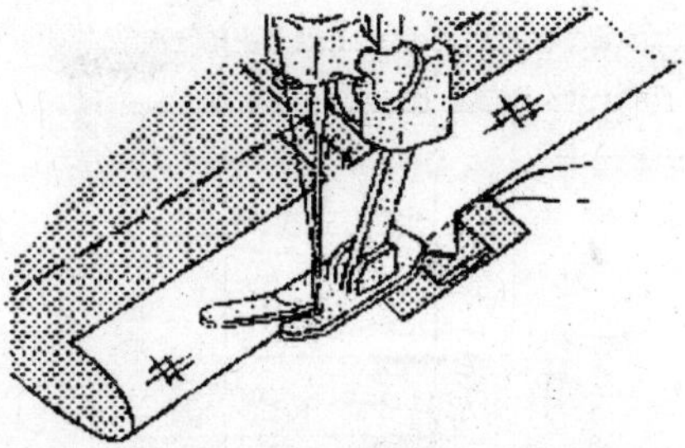

Step 9: Stitch horizontal seam allowance of buttonhole and patch together, just inside the original stitching. Trim patch to 1/4" (6mm). Repeat on other seam.

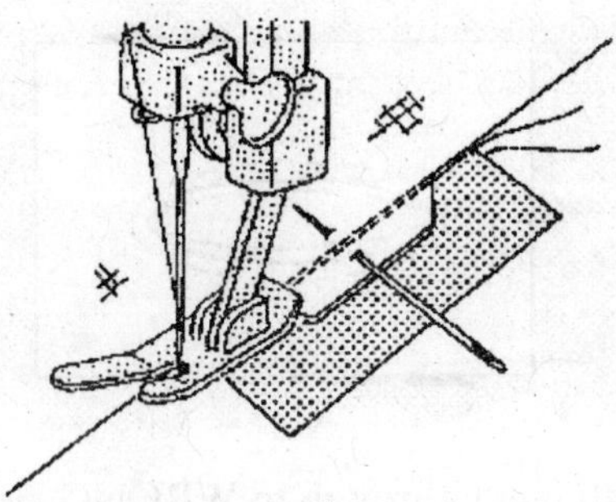

Finishing Your Bound Buttonhole

Step 10: Transfer buttonhole markings to the WRONG (interfaced) side of front facing, and thread-trace markings, same as for bound buttonhole (see step 1). Cut a bias patch of silk organza or lightweight matching fabric 2" (5cm) wide and 1" (2.5cm) longer than the measurement of the buttonhole. Mark a center line horizontally along the patch. With RIGHT sides together, place patch on facing, over thread-traced markings and baste (see step 2).

Step 11: Using small stitches, start stitching at the middle of one side, stitching a scant 1/8" (3mm) from center basting and tapering stitching at ends. Take one stitch across the end, pivot and continue along remaining side. Overlap stitches at starting point. Remove basting

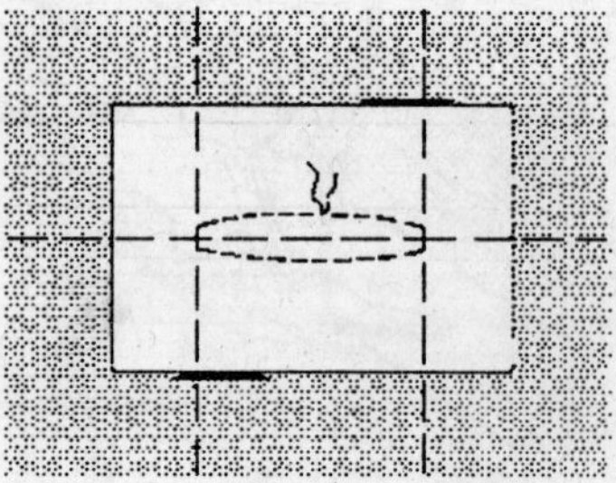

Step 12: Slash between stitching lines, being careful to not clip stitching at ends. Add a drop of seam sealant (such as Fray CheckTM) to ends of opening, and allow to dry.

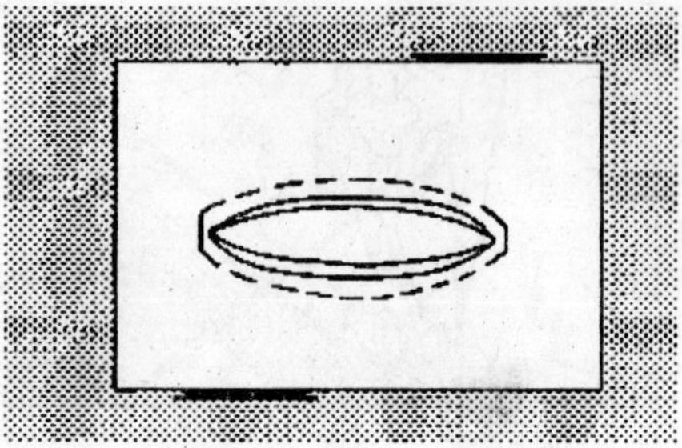

Step 13: Pull patch through to WRONG side. Gently "snap" along the bias of patch to create a smooth slit opening. Press.

TIP: If you prefer, trim edges of patch within 1/4" (6mm) of opening.

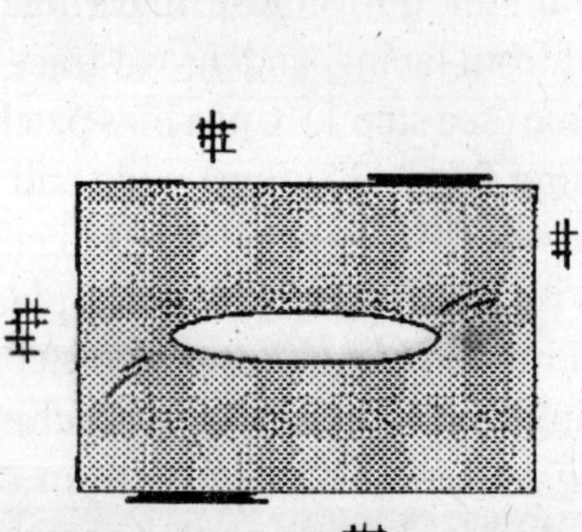

How to Cover Cording to Create Your Own Piping

Choose the cording that you want to cover. Cording is available in a variety of sizes.

Measure the Cording : Using a tape measure

- Lay the cording on the tape measure at the amount of seam allowance you will be using on what you are attaching the piping to.
- Wrap the tape measure around the cording and back to the end of the tape measure.
- Where the end overlays the tape measure is the width of the fabric you will need to cover the coring.

The length of the cording will be the measurement from the area you want to insert the piping. Add a few inches to be sure you will have enough.

Cut Bias Strips : Cut bias strips wide enough to wrap around the cording with the seam allowance.

It is important that this fabric is cut on the bias. Bias will cover the cording smoothly and allows the piping to bend smoothly around corners.

Join the Bias Strips : Match the ends of the bias strips as shown in the photo, using a 1/4" seam allowance, match the long edges of the strip where the 1/4" seam allowance will be sewn.

Press the seam allowance to one side. If you are using very heavy fabric grade the seam allowance(Not eliminating bulk is one of the biggest mistakes when constructing a garment. This gives the appearance of a 'home made' garment. Try the following to eliminate bulk when the seam allowance can not be pressed open or when sewing facings and linings.) to prevent a bulge caused by the seam allowance.

Join enough strips for the length of the cording you need.

Covering the Cording :

- Set up your sewing machine with your zipper foot.
- Place the cording in the center of the bias strip.
- Wrap the bias strip around the cording matching the edges.
- Align the zipper foot so the edge of it is against the cording. Sew the seam allowances together to enclose the cording.

- Allow the machine to feed the fabric under the needle so that you maintain a true bias grain around the cording.

Sew the Piping to the Item : Check that you have the correct seam allowance on the piping for the items seam allowance. Trim or compensate for the seam allowance if necessary.

Using the zipper foot, place the coding on the seam allowance of the item and sew the piping to the item.

Always sew the piping to one layer before joining the layers that will enclose the piping.

Clip where necessary to turn corners.

Ending Continuous Piping : Continuous piping is piping that does not end, as in the edge of a pillow or a cushion cover. There are two methods which are equally acceptable.

***Heavy Piping* :**

- Sew the piping to the item stopping about a half inch before where you started.
- Trim the piping allowing enough fabric to turn under and enclose the beginning of the piping.
- Trim the ends of the cording so that they but up to each other.
- Fold under the fabric to enclose the beginning of the piping and sew in place.

***Narrow Piping* :**

- Sew the beginning of the piping with the end of the piping dipping into the seam allowance.
- Continue to sew joining the piping to the fabric.
- As you approach the starting point, place the ending of the piping over the starting point and dip it into the seam allowance. Allow for a continuous visual line of piping on the outside of the seam allowance.

Ending Piping in a Seam Allowance : When the end of piping will be enclosed in a seam allowance, as in edging a collar, continue sewing the piping to the entire edge. Trim the piping from the seam allowance after all sewing of that area is done to prevent trimming the piping too short.

Enclosing Piping in a Seam : Place the next piece of fabric under the piping, aligning the edges. Use the stitching that is holding the piping on to the first layer as a guide to sew the second layer of fabric. Be sure you sew exactly on the stitching which is holding the piping in place or just beyond the stitching so that the stitching will not be seen from the right side of the item.

Making Fabric Cording : This is one method of making your own fabric cording or rattail cord. Often used as spaghetti straps, drawstrings, button loops, and ties. This works best on fabrics that are a little slippery; if the fabric sticks to itself it will be difficult to turn right side out.

The required materials are:

- 3/4"-1.25" wide strips of fabric, about 2" longer than the finished length you want. A smaller width can be used with thinner, silky fabrics (like satin). A wider width should be used for thicker fabrics. Straps can be cut on the bias for a smoother result, or on the straight-grain to save fabric.
- Thread
- A safety pin
- Patience

Step 1: Fold the fabric strip in half, with the right side to the inside (inside out). Sew 1/4" away from the edge all the way down the length of the strip. Be sure to backstitch to anchor your sewing at the beginning and end of the strip.

Step 2: Pin your safety pin through a good chunk of fabric on the side opposite the seam. If you grab too little fabric, the safety pin will pull out when you turn the fabric.

Step 3: Dive the safety pin into the fabric tube.

Step 4: Turn the cord right side out. Begin to pull the fabric up over the safety pin, and try to work the safety pin through the tube. Once you get it "started" it will be easy to bunch the fabric on the safety pin and pull the safety pin through.

Step 5: Cut the cord into the lengths you want. If you will have a dangling end (like in a tie) finish the end of the cord. You can knot the cord and fray the end as shown in the picture, or tuck the raw edges back into the cord and hand sew the end closed.

11

Cutting and Sewing

Cutting

Cutting is the separation of a physical object, or a portion of a physical object, into two portions, through the application of an acutely directed force. An implement commonly used for cutting is the knife or in medical cases the scalpel. However, any sufficiently sharp object is capable of cutting if it has a hardness sufficiently larger than the object being cut, and if it is applied with sufficient force. Cutting also describes the action of a saw which removes material in the process of cutting.

Cutting is a compressive and shearing phenomenon, and occurs only when the total stress generated by the cutting implement exceeds the ultimate strength of the material of the object being cut. The simplest applicable equation is *stress = force/area*: The stress generated by a cutting implement is directly proportional to the force with which it is applied, and inversely proportional to the area of contact. Hence, the smaller the area (i.e., the sharper the cutting implement), the less force is needed to cut something.

When referring to propagating plants, cutting is one of the methods that can be used. It involves cutting a part of the plant typically a healthy shoot, with sharp and sterile scissors or any other cutting device, and then placing the removed part in water. Some cuttings do not require water. Certain shoots when cut are able to grow when placed in vermiculite or potting soil. However, the former is the easiest to do as most shoots when cut from the main plant need time to grow roots, and then they are able to be transferred to potting soil.

Tools commonly used for cutting :

- Axe
- Blade
- Diamond blade
- Scissors
- Switchblade
- Sword

Rotary Cutting : How to Rotary Cut Fabric Strips for your Quilts

Rotary Cut Fabric Strips : Before you begin rotary cutting, you'll need to square up one end of the fabric, making its edge a 90-degree angle to the fold.

1. Fold the fabric along its length, selvages together. The fold should be straight, with no puckers. Selvages might not be perfectly aligned. Press.

If you are working on a small rotary mat, you may need to fold the fabric again, making it four layers deep. Beginning quilters should stick to one fold, because each new fold makes inaccurate cuts more likely.

2. Place the fabric on your rotary mat with the fold near the bottom edge of the mat and the side to be squared on the left. Align a rotary ruler with the folded edge, its left side near the left edge of the fabric.
3. Place a long rotary ruler to the left of the first ruler, edges flush against each other. Check to make sure horizontal lines on both rulers are parallel to the fold.
4. Remove the right-side ruler. Place your hand on the long ruler to hold it firmly in place and roll a rotary cutter from bottom

to top along the ruler's right edge. Spread your fingers out to hold the ruler securely, but take care to keep them out of the path of the cutter. The fabric's cut edge should now be at a 90-degree angle to the folded edge.

If you are left handed, work from the opposite side of the fabric, placing rulers in mirror-image positions.

Cutting Long Strips : Rotary cut the fabric strips required for your project from the squared-up side. For example, if you need a 3" wide strip, align the 3" mark on the ruler with the left edge of the fabric. Align a horizontal line near the bottom of the ruler with the fold and cut.

Before you make each cut check to be sure that the left and bottom edges of the fabric align with vertical and horizontal markings on the ruler. If they do not, square up the end again before cutting more strips.

It isn't unusual for the leading edge of fabric to become out of square after several cuts, so you're not doing anything wrong if you must square it up occasionally.

Check Strip Accuracy : One way to make sure your edge is square is to open a rotary cut strip to its full width and look at the area near the fold. If the strip has a bend in the middle the fabric's left edge is no longer square.

Rotary Cutting Segments from Strips : Place your rotary ruler near the right end of a cut strip, aligning a horizontal rule with its lower edge. Cut along the right edge of the ruler to square up the end of the strip.

Turn the strip around and cut segments from its left edge in the same way you cut the original fabric strips, aligning the strip with markings on the ruler as required for each shape.

Rotary Cutting Bias Strips : Long bias strips are used to make quilt binding and for appliqué shapes that bend easily into graceful curves—such as flower stems.

To cut long bias strips, align the 45-degree mark on your long rotary ruler with the fold in the bottom of the fabric. Hold the ruler in place and cut along its right edge.

You can cut bias strips along any 45-degree angle of the grain. For long strips, make your first cut near the upper left corner of the folded fabric.

To make the next cut, align the correct line on your ruler with the angled edge of the fabric and the 45-degree line with the fold. Rotary cut along the right side of the ruler again. Continue until you have the number of strips required for your pattern.

Rotary Cutting Tips :

- Always roll the rotary cutter away from your body and follow other rotary cutting safety guidelines.
- If you find it difficult to rotary cut long strips accurately, work with smaller pieces of fabric until you are accustomed to the technique.
- Cut strips along the lengthwise grain instead of the crosswise grain.
- Crisp fabrics are easier to rotary cut than limp fabrics. Stiffen fabrics by spritzing on a bit of spray sizing or starch, then press.
- Attach gripping tabs to the bottom of rotary rulers to help keep them from slipping on fabric. Many types of grippers are available commercially, including clear versions that won't obstruct your view of ruler lines.
- Some quilt patterns tell you to stack strips before rotary cutting shapes from them. The more strips you stack, the less accurate the resulting patches will be. Cut fewer layers at once for best accuracy.

Rotary Cutting Safety Tips for Quilters : Rotary cutting helps you cut fabrics and assemble quilts in record time, but it's a technique that can cause injuries if you don't follow a few basic safety guidelines. Get familiar with rotary safety rules before you ever pick up a rotary cutter.

Using Rotary Cutting Tools : Rotary cutting is one of the most time-saving skills a quilter can master, because it eliminates the need to mark and cut individual patches of fabric. Not only does rotary cutting speed up quilt assembly, if done correctly it will most definitely enhance your accuracy.

Rotary Cutters

Many styles of rotary cutters are available, all with different shapes of handles and a variety of protective sheaths. In general, larger rotary blades allow you to cut more layers of fabric at once—but cutting too many layers at a time can destroy your accuracy, so get plenty of practice before you stack loads of fabrics. Replace the rotary blade when it no longer makes a swift, clean cut through the cloth.

Sewing

Sewing or **stitching** is the fastening of cloth, leather, furs, bark, or other flexible materials, using needle and thread. Its use is nearly universal among human populations and dates back to Paleolithic times (30,000 BC). Sewing predates the weaving of cloth.

Sewing is used primarily to produce clothing and household furnishings such as curtains, bedclothes, upholstery, and table linens. It is also used for sails, bellows, skin boats, banners, and other items shaped out of flexible materials such as canvas and leather.

Most sewing in the industrial world is done by machines. Pieces of a garment are often first tacked together. The machine has a complex set of gears and arms that pierces thread through the layers of the cloth and semi-securely interlocks the thread.

Some people sew clothes for themselves and their families. More often home sewers sew to repair clothes, such as mending a torn seam or replacing a loose button. A person who sews for a living is known as a **seamstress** (from *seams-mistress*) or **seamster** (from *seams-master*), dressmaker, tailor, garment worker, machinist, or sweatshop worker.

"Plain" sewing is done for functional reasons: making or mending clothing or household linens. "Fancy" sewing is primarily decorative, including techniques such as shirring, smocking, embroidery, or quilting.

Sewing is the foundation for many needle arts and crafts, such as applique, canvas work, and patchwork.

While sewing is often seen as a low-skill job, the task of designing good-looking three-dimensional shapes from non-stretching two-dimensional fabric generally requires extensive hands-on knowledge of the design and principles of mathematical manifolds. Flat sheets of fabric with holes and slits cut into the fabric can curve and fold in 3D space in extensively complex ways that require a high level of skill and experience to manipulate into a smooth, ripple-free design. Aligning and orienting patterns printed or woven into the fabric further complicates the design process. Once a clothing designer with these skills has created the initial product, the fabric can then be cut using templates and sewn by manual laborers or machines.

Sewing Tools and Accessories :

- dress form
- dressmaker's or tailor's shears
- measuring tape
- needle
- pattern
- pattern weights
- pin
- pincushion
- rotary cutter
- scissors
- seam ripper
- sewing table
- tailor's chalk
- thread/yarn
- tracing paper
- tracing wheel
- wax, often beeswax
- sewing box

Curtain Heading Styles

Curtain headings are a way of customising and styling blinds and valances so that the overall width of the fabric is reduced, creating a range of pleats or folds.

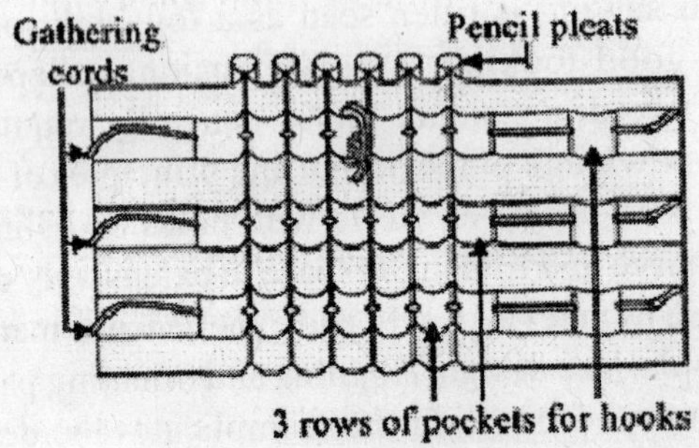

Ready Made Heading Tape

Some headings can be created using various types of tape that is sewn close to the top edge of the window treatment and, by pulling up gathering cords, different effects can be achieved.

Some heading tapes incorporate Velcro at the back so they can be used in conjunction with grip tape that makes it possible to attach top treatments and blinds to battens or pelmet boards.The number of cords and the way the cords are threaded in the heading tape makes the pattern of pleats.

The pleats are sharpest when the cords are drawn up tightly.

There can be 2, 3, or 4 cords for drawing up the pleats.

Most tapes have more than 1 row of pockets for the hooks to adjust the height of the heading.

Box Pleated Headings

Box pleat heading tapes give a very tailored look and drape into deep folds down the full length of the curtains. The pleats butt onto each other with no interval between them. Multiply the track width by 3 when estimating for fabric.

For a more professional finish these headings can be made by hand. Although very time consuming it is well worth the extra effort. As the finished width is critical, but not easily recommend that the width of the pleats and spaces are worked out before the curtains are made so that the width of the flat curtain can be adjusted to suit.

Allow 3 times the window width when estimating for fabric. Suitable for tracks and poles.

Cased Headings

A simple cased heading is used mainly for nets and lightweight fabrics that are not to be opened and closed frequently.

A casing or channel sewn across the top is left open at the ends.

A slender curtain rod or plastic covered spring wire slots through the channel and fits onto hooks or into sockets at the side of the window.

Multiply the rod or wire length by at least 2 when estimating fabric.

Eyelet Headings

For minimalistic look, a narrow rod, pole or wire is threaded through eyelets that are inserted close to the top edge creating deep folds.

Suitable for light to medium weight fabrics.

When estimating for fabric allow twice the rod width.

Goblet Pleated Headings

Goblet pleats make a very elegant heading for longer lengths. Pleats fall from each goblet that can be stuffed with tissue paper or cotton wool to retain their shape. This heading tape has 2 sets of cords and is suitable for medium to heavyweight fabrics.

Multiply the track width by 2 to 2.5 when estimating for fabric. Suitable for tracks or poles.

For a more professional finish these headings can be made by hand. Although very time consuming it is well worth the extra effort. As the finished width is critical, but not easily adjusted I recommend that the width of the pleats and spaces are worked out before the window treatments are made so that the flat width can be adjusted to suit. Allow 2 to 2.5 times the window width when estimating for fabric.

Pencil Pleated Heading Tape

This stiffened tape is available in several depths from 2.5" to 6" (6 cm to 15 cm). A lightweight version is suitable for sheer fabrics.

Some tapes have three rows of pockets for hooks.

Use the top row if the treatment is to hang below a pole. Use a lower row to make the heading stand up the required amount to conceal a track.

Multiply the track width by 2 - 2.5 when estimating for fabric. Suitable for both tracks and poles.

Tab Top Headings

A modern, less formal heading is created with loops of matching or contrasting fabric that are stitched onto the top edge of the treatment omitting the need for heading tape. Suitable for all weights of fabric.

You should also remember to take the length of the tabs into consideration when measuring.

This heading is only suitable for poles.

Measuring Up for Curtains

Measuring up for your window treatments should be one of the first steps, and probably one of the most important, you make. Get this wrong and no matter how well you sew your curtains the desired effect will not be achieved.

For curtains outside the recess the track or pole needs to be wider than the window frame [4" - 6" (10 cms -15 cms)] to allow for stacking the opened curtains. It should be fixed about 4" (10 cms) above the top of the window frame and should be horizontal.

It is advisable to make a plan of each window and transfer all the measurements onto it for future reference. Never presume that 2 windows are exactly the same size even if they look identical.

Take all the measurements in either inches or centimetres. DO NOT MIX THE TWO.

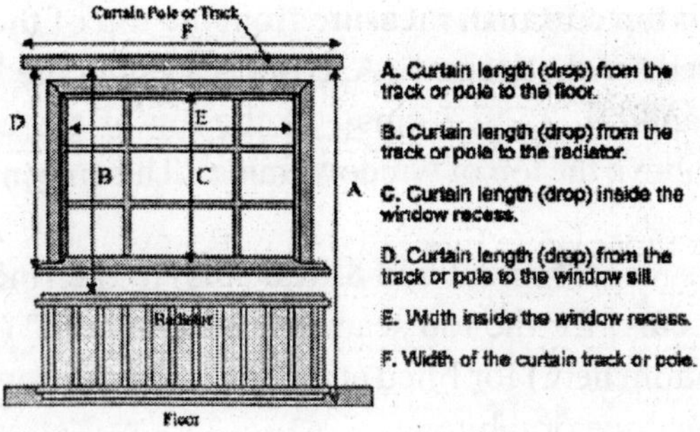

Measuring Curtain Width. Outside the Window Recess :

- Measure the width of the track or pole (F) adding any extra for centre overlapping, normally about 6ins (15 cms).
- Multiply the track width by 1.5, 2 or 2.5 depending on the heading to be used.
- Divide that by the width of the fabric, normally 48" or 54" (122 cms or 137 cms), and round up to the next whole number to give the number of widths required. If an odd number results then one width can be cut in half and added to the outside edge of each curtain.

Measuring Curtain Width. Inside the Window Recess :

- Measure the width of the window recess (E) and multiply by 1.5, 2 or 2.5 depending on the heading to be used.
- Divide that by the width of the fabric, normally 48" or 54" (122 cms or 137 cms), and round up to the next whole number to give the number of widths required. If an odd number results then one width can be cut in half and added to the outside edge of each curtain.

Measuring Curtain Length. Outside the Window Recess :

- For a pole, measure from the bottom of the ring to wherever you wish the curtains to fall (A, B or D).
- For a track, measure from the top of the track to wherever you wish the curtain to fall (A, B or D).

- For tab top curtains, measure from the top of the pole to the required finished length (A, B or D). Deduct the leng:h of the tabs (approx. 3" [7.5 cms]) as the top of the curtain must come above the top of window frame. (This may mean moving the pole.)
- To this measurement add 8" (20 cms) to the finished length (2" (5 cms) for the top seam allowance and 6" (15 cms) for the bottom hem) for lined and unlined curtains with attached heading tape or tab tops.
- For lined and unlined curtains with cased headings add 10" (25 cms) to the finished length (4" (10 cms) for the top seam allowance and 6" (15 cms) for the bottom hem).
- Short curtains should finish 0.5" (12 mm) above the sill or radiator and full length curtains should fall within 1" (2.5 cms) of the floor.
- If patterned material is to be used then each length has to be a full pattern repeat, for example:- With a pattern repeat of 25" and a curtain finished length of 54" you will need 75" for each drop. i.e. 54" + 8" for hems = 62" rounded up to the next multiple of 25" (your pattern repeat). Multiply this amount by the number of widths required and hey presto you have the length of fabric required.
- To calculate the amount of lining required (if used) multiply (the finished length plus hem allowance) by the number of widths required. Hem allowance is usually 6" (15 cms), 2" (5 cms) for the top seam and 4" (15 cms) for the bottom hem. Ignore any adjustments for pattern matching.

Measuring Curtain Length. Inside the Window Recess :

- Measure the length of the window recess (C). Curtains should finish 0.5" (12 mm) above the window sill so they fall correctly.
- To this measurement add 8" (20 cms) to the finished length [2" (5 cms) for the top seam allowance and 6" (15 cms) for the bottom hem] for lined and unlined curtains with attached heading tape or tab tops.

- For lined and unlined curtains with cased headings, add 10" (25 cms) to the finished length [4" (10 cms) for the top seam allowance and 6" (15 cms) for the bottom hem].
- If patterned material is to be used then each length has to be a full pattern repeat, for example:- With a pattern repeat of 25" and a curtain finished length of 54" you will need 75" for each drop. i.e. 54" + 8" for hems = 62" rounded up to the next multiple of 25" (your pattern repeat). Multiply this amount by the number of widths required and hey presto you have the length of fabric required.
- For curtains with heading tape or tab tops, calculate the amount of lining required (if used) by multiplying (finished length + hem allowance) by the number of widths required. Hem allowance is usually 6" (15 cms) [2" (5 cms) for the top seam allowance and 4" (10 cms) for the bottom hem]. Ignore any adjustments for pattern matching.
- For curtains with cased headings, calculate the amount of lining required (if used) by multiplying (finished length + hem allowance) by the number of widths required. Hem allowance is usually 8" (20 cms) [4" (10 cms) for the top seam allowance and 4" (10 cms) for the bottom hem]. Ignore any adjustments for pattern matching.

Inserting a Zip

Place the 2 edges of fabric, right sides together and tack along the seam line.

- Stitch the seam, close to the tacking stitches, leaving an opening where the zip is to be inserted.
- Press the seam allowance open.
- Place the zip, wrong side up, over the tacked section of the seam. The zip can be placed centrally or the teeth can be placed to one side of the seam allowance for a lapped zip.
- Pin and stitch the ends and the sides of the zip, using a zipper foot on your sewing machine, close to the teeth. Note the direction of the stitches on the diagrams.
- Remove the tacking stitches.

Matching Patterned Fabrics

Whatever you are making it is inevitable that sooner or later you will have to join widths of fabric together. This is simple until there is a pattern to contend with. All patterned fabrics have a repeated pattern down their lengths. This is called the "pattern repeat" and is measured from a point in one pattern to the same point in the next pattern.

Most patterned fabrics match horizontally across the width. All the cut lengths have to start at the same point in the pattern so that they match at the seams when they are joined. Therefore a cut length has to be a multiple of the pattern repeat.

Where large patterns are used it is better to have a full pattern at the hemline and a part pattern in the heading where it is less noticeable.

Cutting Out : It is advisable to check the fabric for faults before it is cut because suppliers will not exchange cut lengths.

- Lay the fabric, right side up, on a large flat surface with the top of the pattern at the top.
- Measure the first length along the selvedge, positioning the pattern where it is required and mark with a pin. Do not cut until all the required lengths have been measured to ensure that there is sufficient fabric.
- Measure the required number of pattern repeats from the pin and mark.
- Repeat this until the required numbers of lengths have been marked.
- At each mark cut the fabric at right angles to the selvedge. As each length is cut place a safety pin on the right side at the top to ensure that all the lengths will be joined the correct way up. This is more important when using double-sided patterned fabrics or ones with two-way designs.
- If a half width of fabric is required, fold one length in half, matching the selvedges. Press along the fold.
- Unfold the fabric and cut along the fold. Mark the right side at the top of each half width.

Stitching Widths Together :

- Place one cut length, right side up, on a flat surface.
- Place a second cut length, wrong side up, on top, matching the top raw edges and the selvedges.
- Turn back the seam allowance down the length of the top piece of fabric and press the fold with your finger.
- Adjust the position of the top piece of fabric until the pattern matches at the fold.
- Working down the length of the fabric, release the folded edge and pin both pieces of fabric together, along the fold when the pattern is matched. This fold will be the stitching line.
- Check that the pattern is matched correctly from the right side.
- Machine stitch along the fold.
- Repeat this procedure until the required widths of fabric have been joined together. Add half widths at the outside edge of curtains but always join them at the selvedge.
- Make sure that the bottom edge is straight across all the widths and trim if necessary.
- Measure the length required from the bottom raw edge and cut off the excess fabric where necessary.

Mitring a Hem

This is a neat way of finishing a corner and is less bulky, especially on heavier fabrics.

- Fold up a single or a double hem, mark with a pin where the top fold will come, and press.
- Fold in the side hem, mark with a pin where the inside fold will come, and press.
- Open out the hems.
- Fold in the corner triangle to the pins.
- Turn up a single or a double hem at the lower edge.
- Turn in the side hem.
- Slipstitch across the corner.

Sewing & Soft Furnishing Basic Instructions

Before attempting to make any sort of soft furnishing project it is wise to understand the basic procedures which you will need to master.

Whether it is seams, stitches, measuring or matching patterned fabrics all will need to be mastered.

The links on the menu to the left will take you to the following free instructions to enable you to master and understand some basic sewing techniques.

- Binding is a way of neatening a raw edge using a separate length of fabric. The binding is usually made in a contrasting colour.
- Piping is a method of emphasising a seam or edge and is often used on corners to add strength.
- Curtain Headings are used as a way of customising curtains, blinds and window top treatments by reducing the overall width of the fabric.
- Calculating the amount of fabric required for your curtains or blinds can often be difficult for the novice, in fact I know many professionals who find it difficult. The Online Fabric Estimator will do the hard work for you. This is complimented by the Software Package which is an invaluable and cost saving tool for any home or business user.
- There will no doubt come a time when you need to know how to Insert a Zip. This not only applies to soft furnishing but also to dressmaking.
- When making curtains and blinds it is essential that you know how to Match the Pattern of the Fabric when cutting out. Measuring Up for your curtains and other window treatments is an essential skill which you must learn to master. Mitring is a method used to both neaten and reduce the bulk of fabric on seams and hems.
- Seams are different methods of joining widths of fabric together.
- Have you ever wondered what some of those soft furnishing or sewing terms mean. The Sewing Glossary should help you.

- Stitches are ways of holding pieces of fabric together. Instructions on the most useful and popular are described.

Seams : Throughout the instructions on this web site we have used a variety of seams. Seams are a way of joining fabric together and the more commonly used are illustrated below.

Flat Seam : A simple way of joining 2 pieces of fabric together with a single row of stitches.

- Place the fabric pieces; right sides together, matching the raw edges.
- Pin, tack, and stitch, 0.5" (12 mm) in from the raw edges.
- Work a few stitches in reverse at each end to secure.
- Press the seam allowance open.

French Seam : A self neatening seam that is usually used to join sheer fabrics where no stitches show on the right side.

- With wrong sides together, edges matching, tack and stitch the fabric, taking a 0.25" (6 mm) seam allowance.
- Trim the seam allowance to half its size.
- Refold the fabric with right sides together, to enclose the raw edges.
- Pin and stitch a second seam just big enough to cover the raw edges.

Gathering : Gathering is used to add frills. A frill before it is gathered needs to be at least 1.5 times it's finished length, but making it twice as long gives a much better effect.

When gathering on a sewing machine, use a heavy-duty thread on the bobbin for extra strength and loosen the upper tension slightly.

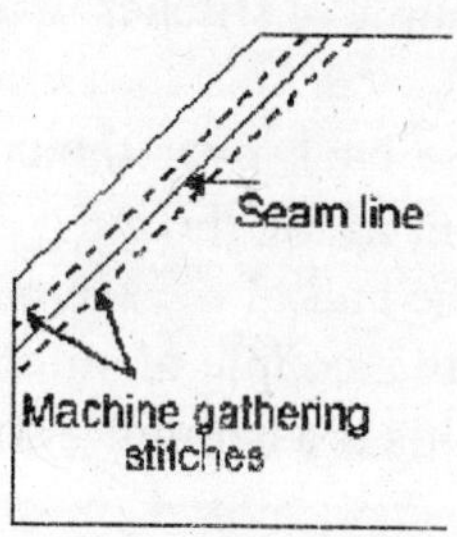

- Work 2 rows of stitches, 0.25" (6 mm) either side of the seam line and knot the threads at one end.

Gently pull the bobbin threads from one end feeding the fabric evenly down the gathers.

- Wind the threads in a figure of eight round a pin at the side when the required length is achieved.
- Instead of machine stitches a small running stitch can be used especially when a long length is to be gathered. It may take longer but there is less chance of the thread breaking.

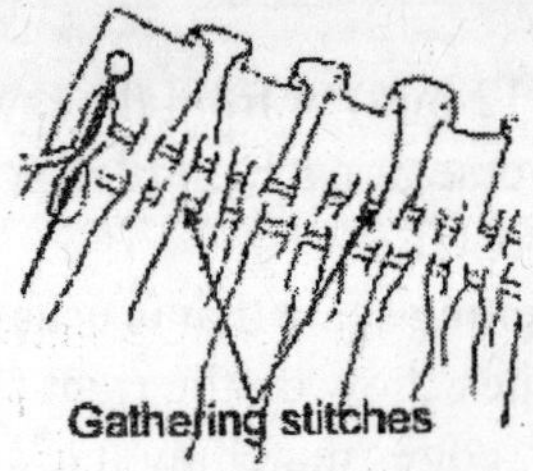

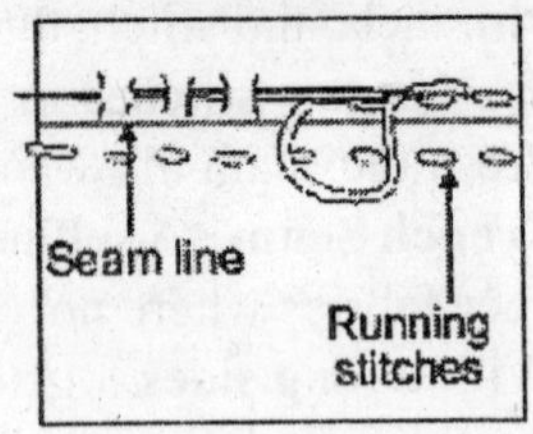

Basic Sewing Techniques

Types of Stitches: Throughout our web site we have indicated the use of different types of stitches which some of you may be unfamiliar with. This page lists a number of the techniques used for these stitches, describes them and with the use of illustrations shows how to sew them

Backstitch: Backstitch is the strongest hand stitch and is used to imitate machine stitches. Work backstitch from right to left.

Begin with a couple of stitches worked on the spot, then take a stitch and a space.

Take the needle back over the space and bring it out the same distance in front of the thread–

–Continue to the end of the seam.

–Fasten off with a couple of stitches on the spot.

Catchstitch: This is a hemming stitch used for bulky fabrics or curved areas–

–Work from right to left.

Fasten the thread to the edge of the folded fabric with a few backstitches. Make a diagonal stitch from right to left, then, with the needle pointing to the left, make a small stitch in the fabric from right to left.

Bring the needle out and make a diagonal stitch from right to left. Do not pull the thread too tight.

Hemming Stitch: Work from right to left with a single thread.

Fasten the thread with a knot inside the hem.

Bring the needle out of the hem and pick up a few threads of flat fabric just above the folded edge.

Make the same stitch through the folded fabric. Work your way along the hem making the stitches as invisible as possible on the right side.

Herringbone Stitch: This stitch neatens a single hem and catches it to the fabric at the same time.

Work from left to right. Secure the thread with a few backstitches.

Make a long diagonal stitch from left to right across the raw edge and back through the flat fabric, about 0.25" (6mm) from the hem edge.

With the needle pointing to the left, make a small stitch in the fabric from right to left.

Bring the needle out and make another long diagonal stitch from left to right so that the threads cross. The stitches should be evenly spaced and the same size.

Oversewing Stitch: Oversewing, or overcasting is the best way to neaten a raw edge by hand to prevent the fabric from fraying.

Relate the length of the stitch to the fabric and how badly it will fray. Begin with a few backstitches.

Make diagonal stitches over the raw edge, spacing them equally and the same length. Be careful not to pull the stitches too tight.

Running Stitch: This stitch is used for seams and for gathering.

Fasten the thread with a few backstitches and work small stitches by passing the needle over and under a few fabric threads and pulling through the fabric. Keep the stitches and spaces as even as possible.

Slipstitch: This stitch is used for holding a folded edge to a flat piece of fabric.

Work from right to left with a single thread fastened with a knot hidden inside the hem. Bring the needle out through the folded edge, pick up a few threads of fabric and then work through the fold again. Slide the needle along, come out of the fold to make the next stitch.

Tacking Stitch: (Basting)

This is used to hold fabric in position while it is being permanently stitched. Work with single or double thread, knotted at the end, and make evenly spaced stitches in and out of the fabric. End a line of tacking with 1 backstitch.

To release tacking stitches, cut off the knot and pull out the thread.

On Cutting and Fitting Garments

Making a Frock. The best way for a novice, is, to get a dress. Fitted (not sewed) at the best mantuamaker's. Then take out a sleeve and rip it to pieces, and cut out a pattern. Then lakt" out half of the waist, it must have a seam in front and cut out a pattern of the back and fore body, both living and outre, part. In cutting the patterns, iron the pieces, smooth, let erec paper be stiff, and then, with a pin, prick holes in the paper to display the gore in front, and the depth of the seams. Thell. with a pen and ink, draw lines from every pinhole, to preserve this mark. Then baste the parts together again, and in doing it, the ullbasted half will serve as a pattern. When this is done, a lady of general ingenuity can cut and fit a dress, by the paltterns thus o btained. If the upper portion of a dress is too tight, the seam under the arm must be let out; and in cutting a dress, there should be an allowance made, to let it out, if necessary, at this seam. The lining of the fore-body must be biassed .

The linings for dresses should be stiffened with cotton or linen. In cutting bias pieces, for trimming; they will not set wel. unless they are exact. ID cutting them. use a long rule and a lead pencil or piece of chalk. Welting-cords should be covered with bias pieces; and it saves time. in many cases, to baste on the welting-cord, at the same time that you cover it. The best way tn but on books and eyes. is to sew them on double broad tape. and then sew tbis on the frock.lining. They can then be moved readily, and do not display where seved on.

In cutting a sleeve. double 'it biased. The skirts of dresses look bodly. if not full and in putting on lining. at the bottom be careful to have it a very little fuller than the dress, or it will shrink, and look badly. All thin silks look much better with lining, and last much longer, as do aprons, also. In putting in a lining to a dress. baste it on each separate breadth. and it looks much better than to have if fastened only at the bottom. Make notches in selvage to prohibit it from drawing up the breadth. Dresses which are to be washed, should not be lined.

Silk dresses are very expensive; and none but the wealthy 'should use them, barring for a full dress. Any lady will look properly dressed. for almost any opportunity. in the beautiful prints which are now to be obtained, and which wash and we-are so well, as to make them the most economical dresses, both for Winter and Summer .

Figured silks, do not usually wear well, if the figure is large and iatin-like black and plain coloured silks can be tested by getting samples, and making creases in them; fold the creases in a bunch, and rub them against a rough surface, of moreon or carpeting. Those that are poor, will soon wear off, at the creases. English satin reps of the best quality. bouble Florence. or Marceline, gros d' Afrique, and Turkish satin. are the best silks for wear .

Plaids are becoming, for tall women, as they shorten the appearance of the figure. Stripes are becoming to a large person. as they reduce the apparont size. Pale persons should not wear blue or green, and brunettes should not wear any lig..t delicate colours, barring shades of buff, fawn, or straw colour. Pearl white is not good for any complexion. Dead white and black are becoming to almost all persons. It is best to try colours, by candle-light, for evening dresses; as some colours, which look very handsome in the daylight, are very homely when seen by cancle-light. Be careful, in cutting dresses, to have the figures run in the right direction, and to have all the parts agree. When there is a right and wrong side, and a figure also to be matched, much care is essential to prohibit mistakes. Never cut a dress low in the neck, as it exhibits that a woman is not properly instructed in the rules of modesty and decorum, or that she has not sense enough to regard them. Never be in haste to be first in a fashion, and never go to the: extremes.

To Cut and Make Shirts and Chemises

In purchasing linen, look for that which has a round close thread. and is perfectly White; for if it he not white at first, it will never afterwards become so. Much that is called linen at the shops half cotton, and does not wear So well as cotton alone. Cheap linens are usually of this kind. It is not easy to discover which are all linen; but the best way is to find a lot, presumed to be good, take a sample, wash it and ravel it. If this be good, the rest of the same lot will perhaps be so. If you cannot do this, draw a thread, each way, and if both appear equally strong, it is possibly all linen. Linen and cotton must be put in clean water, and boi led, to get out the starch, and then ironed. A long piece of linea, a yard wide, wil', with care aud calculation, make eight ~birts. In cutting it, take a shirt of the right size, for a model, and by it measure the length of sevel bodies, not cutting any but the last. Cut off from this long piece all that is required to go into the bodies, and from the parochiat strip cut out the binders or linings, gussets, and other parts, measuring by the pattern. Then cut out the sleeves, from the original piece: and when seven shirts are cut out, calculate whether there is enough for the eighth, before starting to cut it. Take the model-shirt as a guide, in fitting and basting.

Bosom-pieces, false collars, must be cut and fitted by a pattern which suits the person for whom the articles are designed. Gentlemen's night-shirts are made like other shirts barring that they are longer .

In cutting chemiset, if the cotton or linen is a yard wide cut off small half gores, at top of the breadths, and set them on the bottom. Use a long rule and pencil, in cutting gores.

In cutting cotton which is rather wide, a seam can be saved

by cutting out two at once, in this manner. Cut off three breadths, and, with a long rule and a pencil, mark and cut off the gores thus from one breadth, cut off two gores, the whole length, each gore one fourth of the breadth at the bottom and tapering off to a point at the top. The other two breadth are to have a gore cut off from each which is one fourth widC' at top and two fourths at bottom. Arrange these peces right and they will make two cbemises, one having four seams, and the other three. This is a much easier way of cutting, than sewing the three breadths together, in bag fashion, as is often done. The biased, or goring seams, must always be felled. The sleeves and neck can be cut in conformily with taste of the wearer, by another chemise for a pattern. There should be a lining around the armholes, and stays at all corners. Five yard of yard width, will make two chemises.

Old silk dresses, quilted for skirts, are very serviceable. White flannel is soiled so readily, and shrinks so much in washing, that it is a good plan to colour it a light dove-colour according to a earlier receipt. Cotton flannel, dyed thus, is also good for common skirts. In making up flannel, back-stitcb and run the seams, and then cross-stitch them open. Nice flannel, for infants, can be ornamented, with very little expense of time, by turning up the hem, on the right side, and making a little vine at the edge, with saddler's silk. The ~titch of the vine is a change of buttonhole.stitch

Prolong nightgowns are best cut a little goring. It needs five yards, for a long nightgown, and two and a half for a short one. Linen nightcaps wear longer than, and do not turn yellow as cotton ones do. They should be ruffled with linen, as cotton borders will not last so long as the cap.

A double-quited wrapper is a great comfort, in case of sickness. It may be made of two old dresses. It should not be cut rather like a gentleman's study-gown. having no collects or plaits, but sufficient enough to slip off and on with ease. A double gown, of calico, is also very useful. Most of articles of dress, for grown persons or children, need patterns.

Bedding. The best beds, are thick hair mattrasses, which, for persons in health, are good for Winter as well as Summer use.

Mattrasses are also made of husks, dried and drawn into shreds. They are also made of alternate layers of cotton and moss. The most beneficial sheeting, is the Russian; which will last three times as long as any other. It is never perfectly white. Unbleached cotton is good for Winter. It is poor economy to make narraw and short sheet, as children and domestics will in cessantly slip them off, and soil the bedtick and bolster. They should be three yards long, and two and a half wide, so as to tuck in all around. All bed-Iinen should be marked and numbered, SI) that a bed can always be made appropriately, and all missing articles be known.

MENDING

Silk dresses will labt much longer by ripping out the sleeves when thing and changing arms, and also by chahging the breadths of the skirt. Tumbled black silk, which is old and rusty, should be dipped in water, and after being drained for a few minutes, should be ironed, without squeezing or pressing. Cold tea is better than water. Sheets, when worn thin in the middle, should be ripped, and the other edges sewed together. Window-curtains last much longer, if lined, as the sun fades and rots them. Broadcloth should be cut with reference to the way the nap runs. When pantaloons are thin, it is best to newly seat them, cutting the piece inserted in a curve, as corners are not easy to fit. When the knees are thin, it is a case of domestic surgery, which demalds amputation. This is formed, by cutting off both legs, some distance above the knees and then changing the legs. rake care to cut them off properly the same length, or in the exchange they will not fit. This method brings the worn spot under the kness, and the seam looks much better than a patch and darn. Hose can be curtailed when the feet are worn. Take an old sale stocking, and cut it up for a pattern. Make the heel short. In sewing, turn each edge, and run it down, and then sew over the edges. This is better than to stitch and then cross-stitch. Run thin places in stocking, and it will save darning a hole. If shoes are worn through on the sides, in the upper-Ieather, slip pieces of broadcloth under, and sew them around the holes.

Sewing & Soft Furnishing Glossary

Acorn : A decorative brass or wooden "handle" which slips onto the cords of a blind to cover the knotted ends.

Arched Valance : A valance which is shaped at the lower edge, the sides being longer than the centre.

Austrian Blind : A gathered blind made longer than necessary. The extra fabric forms ruching at the bottom.

Banding : Another word for binding.

Batten : A length of wood to attach a blind, pelmet or valance to a window frame.

Bed Valance : A skirt covering the divan part of a bed.

Bias : A diagonal line across the fabric.

Bias Binding : Strips of fabric cut on the bias.

Binding : A way of neatening a raw edge with a separate length of fabric.

Bishop's Sleeve Curtains : Curtains that are made longer than necessary, tied tightly, pulled up and flounced over the tied section.

Blind : A single curtain with a fixed heading which pulls up from the bottom.

Bound Edge : A way of neatening a raw edge using bias binding.

Box Pleats : A row of folds in alternate directions. The extra fabric in the folds can be taken either to the front or the back for the desired effect.

Braid:A decorative trim.

Buckram: Cotton or jute fabric for stiffening, sometimes come impregnated with glue.

Bullion:A thick twisted fringe used for decoration.

Cafe Curtains:A curtain made to fit the bottom half of a window only, to give privacy but to let in light.

Cased Heading:A channel at the top of the curtain to thread a curtain rod through.

Casing:A stitched channel between two pieces of fabric to hold either a length of dowelling or a curtain rod.

Chain Weights:A continuous chain of small heavy beads covered in a cotton casing used for lightweight fabrics.

Cleat:A two pronged hook which is fixed to one side of the window frame to secure the cords when a blind is pulled up.

Combination Rods: Two or three curtain tracks sharing one set of brackets. Used for a layered look of curtains and top treatments.

Contrast Lining:A coloured fabric used as a lining when parts of it will show from the front.

Cording: Another word for Piping.

Cornice: Another word for a Pelmet.

Another word for a Cafe Curtain.

Curtains: Usually two rectangles of fabric hung from a track or pole to decorate a window and give privacy.

Cut Width: The width of fabric needed including seams or hems.

Double Fullness: When each of a pair of curtains are the measured width of the window enabling curtains to drape in folds.

Double Hem: Folding the fabric over twice in equal amounts i.e. a 2" double hem would need 4" of fabric.

Dowelling:A circular or oval length of wood or plastic attached to the back of a blind to keep the fabric flat.

Drapes: Another name for curtains.

Dust Skirt / Ruffles: Another name for a bed valance.

Envelope Curtains: Casual curtains that do not pull back. The bottom inside corners are hooked back to let light in.

Face Fabric: The main fabric that is used for the front. Sometimes called decorator or self fabric.

Festoon Blind: Often confused with Austrian blinds the difference being that a Festoon blind is ruched from top to bottom.

Finial:Decorative ends of a curtain pole.

Finished Width: The actual width after the treatment is finished and all allowances have been utilised.

Frill:A longer length of fabric gathered or pleated onto an edge for decoration.

Fullness Ratio: This is the ratio of fabric width to the width of the window. Curtains are usually at least twice the window width.

Header: The extra fabric above a cased heading which forms a frill.

Heading Tape: A wide woven tape incorporating pockets for curtains hooks and gathering cords.

Hold Backs: Decorative brass hooks or mushrooms fixed onto the wall to hold curtains back.

Interlining: A soft fluffy fabric placed between the face fabric and lining to help insulation.

Inverted Pleat: A flat pleat with the extra fabric to the wrong side.

Jabot: The tail section of Swags and Tails.

Kick Pleats: Similar to box pleats but the folds do not butt together at the back, making the pleats further apart.

Knife Pleats: A row of folds all in the same direction.

Lambrequin: A pelmet which extends down the side of the window.

Leading Edges: The central vertical edges of a pair of curtains.

Mitring: A way of folding the excess seam allowance to achieve a less bulky, sharp corner.

Nap: A one way direction of texture on a fabric such as velvet or corduroy. When using fabric with a nap all pieces must be cut with the nap in the same direction.

Pattern Repeat: The amount one pattern is duplicated down the length of the fabric. Pattern repeat is one full pattern.

Pelform: A double sided sticky card especially made for pelmets and tie backs, printed with various edge patterns.

Pelmet: A decorative way of concealing the top of curtains and curtain tracks. Usually a flat shaped panel which can be painted or covered with fabric.

Pelmet Board: A horizontal wooden shelf from which a pelmet or valance is hung.

Pillow Sham: A decorative pillow covering used during the daytime.

Piping Cord: A fabric covered cord inserted to accentuate or decorate a seam.

Puddled Curtains: Curtains made longer than necessary to allow them to puddle onto the floor.

Railroading: Using fabric horizontally rather than vertically. Fabric without a directional design or nap can be railroaded to avoid seams in long lengths such as bed valances.

Return: The sides of the window treatment that project from the wall.

Ruffle: Another name for a Frill.

Scarf: A length of fabric which is neatened and draped across the top of a window treatment.

Seam: A way of stitching two pieces of fabric together.

Seam Allowance: The amount of extra fabric added to make a seam.

Selvedge: The woven outside edge of the fabric.

Stackback: The amount of space taken up by the curtains or draperies when they are open.

Tie Backs: Stiffened shapes of fabric hooked onto the wall to hold curtains back.

Valance: A gathered, and sometimes shaped, mini curtain hung from a pelmet board to conceal the top of curtains or a curtain track.

Warp: The threads that run down the length of a woven fabric.

Weft: The threads that run across a woven fabric.

Welt: A strip of fabric sewn between the two sides of a cushion to give it more depth.

12

Sewing Tools and Equipments

Sewing tools aren't just the sewing box that Grandma had. There are all kinds of tools and equipment that our grandmother's never could have envisioned. Having the right tool for the job makes accuracy and achieving professional results possible.

1. Hand Sewing Tools
2. Sewing Machines
3. Machine Embroidery

Hand Sewing Tools

Hand sewing tools are more than Grandma's sewing box. They may not all fit in a sewing kit but they will contribute to accurate

- Create a Sewing Kit
- Photo Gallery of Sewing Tools
- Sewing Tape Measures
- Sewing Gauges

- Pincushions and Magnetic Pin Holders
- Seam Rippers
- Needle Threader
- Hand Sewing Needles
- Bent-handle Dressmaker Shears
- Pressing Tools for Professional Sewing Results
- Pinking Shears
- Bodkin
- Sewing Rotary Cutting Mats
- Fabric Rotary Cutters
- Rotary Rulers
- Sewing Needle Threaders
- How to Use a Needle Threader to Thread a Sewing Needle
- Is There a Difference In Threads? You Be the Judge!

Create a sewing kit

Most kits on the market leave something out or have an item that is inferior quality which will lead to more frustration then money you may have saved.

A storage container is a great place to start.

Items to include in a sewing kit:

- *2 pair of scissors*- an inexpensive pair of paper scissors and a good quality pair of fabric shears. By including both pair of scissors, you are insuring that the fabric scissors will last a long time.
- *Straight pins & Pin Cushion* - many people, find the quilting pins with coloured heads, easier to handle. Visit a notions department and see the vast assortment available on today's market. A magnetic pin holder is also available in most notions departments.
- *A tape measure* -A sewing tape measure is a non-stretchable flexible strip usually marked with inches on one side and centimeters on the other.
- *Tracing wheel* - there are smooth wheels and spoked wheels on the market. Each has it's place and are great notions for any sewer to have on hand.

• *Dressmaker's carbon* - this is a special carbon paper that is available in combination packs or for light fabric or for dark fabric. This is available in department stores that have a fabric department or at fabric stores.

• *A gauge* - This is a great tool for making accurate measurements and for tracing straight lines such as darts.

• *A seam ripper*- These are available in many sizes and some are glorified. All new seam rippers are sharp and precise.

• *Hand sewing needles* - a nice assortment to fit various sewing needs Thread - a variety of basic colours in a good quality thread. This is one place the skimping is not worth while.

Scissor - Accurate cutting for any given sewing task helps maintain accuracy as you precede... the key to having pieces fit together.

These shears have straight blades with a handle that is off-set at an angle allowing the lower blade to stay flat on a cutting surface.

The design of the handle allows the bottom blade to rest on the flat surface below the fabric without your hand lifting the fabric off the flat surface. The fabric remaining flat and smooth increases the likelihood that your cutting will be accurate.

***Use*:** Bent-handle dressmaker shears are used for the cutting out of patterns and sewing projects. Sharp shears are the key to preventing hand fatigue and accurate cutting along pattern lines.

***Care & Maintenance*:**

- Use them only for fabric! A sewing box or area secure from other family members is best.
- Keep the blades clean and oil them occasionally.
- Do not over extend the blades by trying to cut through too many layers at a time. Sharp shears will cut with no effort. If you have to force the shears consider cutting less layers at a time.
- They should not be the scissors that are kept in the junk drawer or kitchen drawer. Using fabric scissors on anything other than fabric will dull them, causing uneven cutting, shredding of fabric and hand fatigue.

***Available Options*:** These shears are available in simple to deluxe versions. Simple versions have ordinary blades and handles. Cushion grips of varying forms are also available. Spring-loaded versions which help hands with problems such as carpal tunnel and arthritis are also available.

Straight Pin and Pin Cushion :A pincushion holds the straight pins and needles while you are working to prevent accidents. Magnetic pincushions are extremely helpful if you tend to toss pins on your work surface. The magnet helps in keeping pins located or corralled and can pick up a spill in seconds. The drawback is that if the pins are left on the magnet for an extended period of time, they will become magnetized and eventually you are apt to find them on your scissor blades and other places that you do not want them. Because they will "collect" pins, if you like keeping all your types of pins separated, this may not be the tool for you.

The standard tomato pincushion seems to have been around forever. A good pincushion is filled with sawdust and wool roving. The wool roving contains lanolin and prevents the pins from rusting.

The strawberry that is attached to most pincushions (but also available separately) is filled with fine sand or emery. The abrasive action of these fillers, remove dirt and rust, keeping the pin or needle sharp and smooth.

You must take the time to place the pins in the tomato type of pincushion for it to be useful but it does more to maintain your pins than a magnetic type of pin holder.

***Use* :** Pincushions may stay in your sewing area and just be handy by the sewing machine.

***Care & Maintenance*:** With magnetic pin holders you must be cautions around electronic equipment and data storage devices to prevent the magnet from damaging them.

Prevent tomato type pincushions from getting holes. The spill out ic very messy to clean up.

***Available Options*:** There are pincushions that have been in sewing boxes for generations and many new options available. Some can be worn on a wrist to have handy when pinning alterations or hand sewing. Small decorative options are also available. There are also elaborate decorative pincushions available at many sewing stores (not pictured here).

***Expense*:** Prices vary widely depending on the quality of the filling and the novelty of the type of pincushion. Look for a tight weave on any fabric type of pincushion to prevent leaking of the contents.

Tape Measure : A tape measure is a must-have tool sewing. Having more than one and carrying one in your purse not uncommon!

***Description*:** A sewing tape measure is a non-stretc flexible strip usually marked with inches on one side and centimeters on the other.

***Use*:** Taking body measurements is the most common use a tape measure but it can be used for any measuring task. Becaus it is flexible but firm, it can stand on its edge and measure the distance around an item, such as a circle.

***Care & Maintenance*:** Tape measures require very little maintenance. Simply keeping it clean and using it for measuring will preserve a tape measure for years. If the kids were to get a hold of it for a game of tug-of-war, be sure to check the accuracy of the markings just in case it did manage to stretch.

***Available Options*:** The size of the markings, the width of the tape measure and auto retractable are a few of the options available.

***Expense*:** Tape measures are a very inexpensive sewing tool. ometimes they are even given away at store openings or other romotional opportunities. A retractable tape measure for your purse is not an expensive investment and you will find many uses for it when you are out and about.

Tracing Tools : Tracing Wheels are available in a variety of forms. Wheels are smooth or spoked. Some have extra features such as a smooth area for pressure marking. there are smooth wheels and spoked wheels on the market. Each has it's place and are great notions for any sewer to have on hand.

Dressmaker's Carbon : Dressmaker's Carbon is available in non-wax, wax and vanishing forms. They are all a sheet for of transferring material. this is a special carbon paper that is available in combination packs or for light fabric or for dark fabric. This is available in department stores that have a fabric department or at fabric stores.

Sewing Gauge : Sewing gauges are hand tools for measuring small areas as you are sewing.

***Description*:** Gauges are available in a variety of forms. The most common to have is a six inch aluminum ruler with a slider. More elaborate forms of this gauge are a bit longer allowing you to full utilize six inches while working with the gauge.

Small, almost jigsaw puzzle shaped gauges have markings for the most common sewing measurements that are either seam allowances, turned under or hemmed.

***Use*:** Although gauges are not a necessary tool, they are inexpensive and compact. Gauges are handy to have for all hand sewing, marking alterations, hemming, checking the width of your seam allowances and more.

The straight edge type of gauge is also useful as a straight edge when drawing alteration lines.

***Care & Maintenance*:** Do not allow your metal gauges to get bent. Just like a car fender, once it gets bent, it is not going to go back to the same size or shape it was before, causing the measuring aspect of the gauge to be distorted.

***Available Options*:** Some are small with commonly used seam allowanced gauged around the tool. Some have a sliding marker to keep your measurements uniform. More elaborate gauges have marking chalk build into the tool.

***Expense*:** Most gauges can be found for less than $5. Watch for clearance sales to pick up bargains...some can be found for less than a dollar.

Sewing Seam Rippers : An oddly shaped, inexpensive, sharp device that makes removing stitches possible with less likelihood of ruining fabric.

***Description*:** A small sewing notion that usually has two points, one sharp and one that is protected so that it will not accidentally penetrate and rip fabric.

Use: A seam ripper is used to remove unwanted stitches. The fine tip of a seam ripper allows you to pick out single threads decreasing the likelihood of cutting the fabric that the stitches are attached too.

Care & Maintenance: Keep your seam ripper clean and use it only for sewing tasks to keep it sharp. Keep it covered when you are not using it to prevent injuries and damage.

Available Options: The seam ripper shown here is the most commonly used for sewing. Larger, heavier seam rippers are available for heavy removal jobs or cutting through heavy threads.

Expense: Seam rippers are an inexpensive sewing notion. They are worth replacing as soon as they start to dull.

Hand Sewing Needles : Make the correct choice when choosing a hand sewing needle to eliminate damaging fabric and make the task easier.

Description: Hand sewing needles are available in varying sizes with varying points. They guide the thread through fabric when you are hand sewing.

Use: Hand sewing needles are used in all types of hand sewing. The most commonly used hand sewing needles are called sharps. Sharps have a medium length (in comparison with all available needles), have a round eye for the thread and are suitable for almost any fabric.

Specialty needles are rarely used. A curved needle is not for sewing circles. It would commonly be used for hand sewing upholstery, when a straight needle would not fit.

Thick heavy needles would be used on fibres that are thick and difficult to sew through. The finer the fabric, the finer the needle you would choose.

Care & Maintenance: The strawberry helps to keep needles sharp and smooth. Read more about the strawberry in pincushions.

Available Options: Hand sewing needles come in a wide variety of shapes and forms. Test your choice of sewing needle on an inside seam to be sure it will not be making holes in the fabric.

Expense: Hand Sewing needles are a very inexpensive sewing notion. They are widely available. When sewing expensive and delicate fabric, do not scrape-by with inexpensive needles that could lose their coating and snag the fabric, instead buy a brand name at a fabric store.

Optional Items

- Rotary Cutter, cutting pad and Quilting rulers - These are fantastic to have on hand when cutting straight lines.
- Thread Snips - These are a very handy devise to have next you your sewing machine. They are available in a variety of colours.
- A Sewing Magazine Gift Subscription and recommend Sew News and Threads
- Sewing Books
- A notebook with page protectors and print outs of the many, many free on line projects.
- Hardware Bin organizer Boxes - available in most hardware stores. Some are able to be hung on the wall and make a great organizer for machine feet and tools.
- Thread racks - make your own or explore the many available on the market at most fabric stores.
- Light weight fusible interfacing - yards of it are great to have on hand. This makes a great backing for tissue patterns that become a favorite, to keep them strong and durable.

It can be difficult to shop for a tool when you are not even sure what it looks like. You may not even be aware of some tools that will make a sewing technique easier or give you better finished results. Follow the gallery to see sewing tools. Click the image of any tool for in-depth information about the tool.

Sewing Rotary Cutting Mats : Rotary cutting mats are a basic tool that you must have to use a rotary cutter.

***Description*:** A rotary cutting pad is available in a variety of sizes. It is a self-healing mat that is not damaged by a rotary cutter and it does not damage the rotary cutter.

***Use*:** A rotary cutting pad is is used as a flat surface to cut fabric with rotary tools

Care & Maintenance: They are self healing to any cuts made by the rotary cutting tool. Do not damage your mat by using it with a heavy cutting blade such as a knife or razor tool.

Always use a cutting mat on a hard flat surface! A cut through the mat will not heal.

Fabric Rotary Cutters : A rotary cutter is a fast way to cut accurate straight lines when used in conjunction with a rotary pad and rotary rulers.

Description: Rotary cutters are round blades with a handle and protective cover for the blade.

Use: Rotary cutters are used in conjunction with rotary pads and rotary rulers.

Care & Maintenance: The cover should always cover the blade when the cutter is not being used. Just as you would not use fabric scissors on paper do not use a fabric rotary blade on paper, to prevent it from becoming dull.

Available Options: Rotary cutters are available with different handles and protective devices. Always lock the blade cover when small children are in the area and the cutter is not in your hands.

The blade of thei Cutter is always covered unless you are applying pressure in the process of cutting. There is a lock to hold the blade cover in place when the tool is not in use or being transported. The handle has ridges which are comfortable to hold and prevent your hand from sliding.

Rotary Rulers : Rotary rulers are a great tool to assure that you are getting squares and angles that are perfect.

Description: Omnigrid rulers feature 1" through 1/8" markings as well as 30, 45 and 60 degree angle markings. The small black dots make aligning the ruler easy and accurate.

Use: Rotary rulers can be used alone as a measuring devise or used in conjunction with rotary mats and rotary cutters.

Care & Maintenance: Rotary rulers are clear plastic can can be broken.

Available Options: They are available in a variety of sizes. The most commonly used size is 6" X 24".

Bent-handle Dressmaker Shears : Accurate cutting for any given sewing task helps maintain accuracy as you precede... the key to having pieces fit together.

***Description*:** These shears have straight blades with a handle that is off-set at an angle allowing the lower blade to stay flat on a cutting surface.

The design of the handle allows the bottom blade to rest on the flat surface below the fabric without your hand lifting the fabric off the flat surface. The fabric remaining flat and smooth increases the likelihood that your cutting will be accurate.

***Use*:** Bent-handle dressmaker shears are used for the cutting out of patterns and sewing projects. Sharp shears are the key to preventing hand fatigue and accurate cutting along pattern lines.

***Care & Maintenance*:**

- Use them only for fabric! A sewing box or area secure from other family members is best.
- Keep the blades clean and oil them occasionally.
- Do not over extend the blades by trying to cut through too many layers at a time. Sharp shears will cut with no effort. If you have to force the shears consider cutting less layers at a time.
- They should not be the scissors that are kept in the junk drawer or kitchen drawer. Using fabric scissors on anything other than fabric will dull them, causing uneven cutting, shredding of fabric and hand fatigue.

***Available Options*:** These shears are available in simple to deluxe versions. Simple versions have ordinary blades and handles. Cushion grips of varying forms are also available. Spring-loaded versions which help hands with problems such as carpal tunnel and arthritis are also available.

***Expense*:** The amount you spend is a matter of choice. I have seen this style shear for a dollar and for over fifty dollars. The more you spend, the more comfortable the handle is apt to be and the longer the blade will stay sharp, as long as it is used for only fabric.

Pressing Tools for Professional Sewing Results : The correct tools to do any sewing job will help you achieve the best results possible.

***Description*:** The most basic sewing tools are an iron and ironing board.

A press cloth is a piece of fabric used when applying more heat than you usually would when ironing a fabric. They provide a barrier to protect your fabric from direct contact with a hot iron.

Hams and sleeve rolls are very firmly packed pillow type curved objects with two different types of fabric surfaces.

Presser bars are sticks of wood with a flattened end. They work well to hand press small seams on craft objects or when an area is totally impossible to reach with an iron.

Small ironing boards are miniature versions of the ironing board everyone has in their closet.

***Use*:** A press cloth is essential to prevent scorching your fabric when you are applying more heat than you would if you were just ironing out wrinkles. They are available in different weights. A see-through press cloth allows you to see the pressing task through the press cloth. If you do not have one, a piece of muslin will do but it is an item worth adding to your sewing tools.

Hams and sleeve rolls are wonderful for pressing curves and seams that are in a tube area without creasing another area of the garment. Similar results can be achieved with a tightly rolled towel. It is worth saving up for a pressing ham and sleeve roll if you will be sewing clothing, baby clothes or doll clothes.

A small ironing board can achieve pressing the inner seam of some tubes. Because it has a stand, it is not as flexible to use as a sleeve roll.

***Care & Maintenance*:** Always read the information that comes with any tool to prolong the life of the tool.

Pinking Shears : Pinking shears allow you to cut a simple seam finish.

***Description*:** Pinking Shears have jagged blades that fit together to cut a saw-tooth type edge.

Use: A pinked edge to many tightly woven or non-fraying fabrics is sufficient as a seam finish. This is especially handy when working with light-weight fabrics that a sewn seam finish will add to much thread or weight to the seam.

On non-fraying fabric such as fleece using pinking shears helps reduce a blunt edge from showing through when pressing a seam and adds a finished look to the fabric.

Care & Maintenance:

- Use them only for fabric to prevent dulling the blades! A sewing box or area secure from other family members is best.
- Keep the blades clean and oil them occasionally.
- Do not over extend the blades by trying to cut through too many layers at a time. Sharp shears will cut with no effort. If you have to force the shears consider cutting fewer layers at a time.
- They should not be the scissors that are kept in the junk drawer or kitchen drawer. Using fabric scissors on anything other than fabric will dull them, causing uneven cutting, shredding of fabric and hand fatigue.

Available Options: Pinking scissors are available in five and a half inch to ten and a half inch lengths. The most common choice is seven and a half inch.

Expense: The amount you spend is a matter of choice. Look for strong blades on this type of shear. When you are going to be using them on heavy fabric, it is best to invest in a good pair.

Bodkin : A bodkin is a handy tool to have but not a *must-have* item. It is a tool that helps replace drawstrings and other items that are enclosed in a casing.

Description: A bodkin looks like an over-sized sewing needle with a blunt point and a large eye.

Use: A bodkin is designed to pull an item (such as elastic, ribbon, cording or drawstring) through a casing. The blunt point, sometimes a ball point, prevents the bodkin from piercing the fabric casing, while guiding the pulled through item, through the casing.

If you do not have a bodkin a safety pin can be used to pull the item through a casing. It is just a bit more difficult to grasp and guide than a bodkin.

***Care & Maintenance*:** No special care is required.

***Available Options*:** A bodkin can have various textures. The varying textures allow for choices when working with different fabric types. For example; when trying to place a drawstring through a slippery fabric, It would be more apt to grab the textured bodkin so that It could gain a better grip on the bodkin as it passed through the casing.

A bodkin may also resemble large tweezers. You grip the item being pulled in the tweezers and slide a ring down over the tweezers to hold it in place, so it can be threaded through the casing. This is very helpful for an item that is too bulky to be doubled over itself when it is in the eye type of bodkin.

***Expense*:** A bodkin typically costs under two dollars. As new products are designed with more details, the prices will increase to meet new innovations.

How to Use a Needle Threader : The Needle threader consists of two parts : The handle and the wire. The end of the wire that is away from the holder is folded. Place the folded wire of the needle threader through the eye of the sewing needle. Place the thread into the wire loop of the needle threader. Holding on to the body of the needle threader, pull the wire part back through the eye of the needle, stopping when you have one thread in the eye of the needle. Remove the thread from the wire part of the needle threader. Knot the thread and sew.

Sewing Tools : A sewing machine is probably the most expensive investment you will make in sewing. Having a machine that functions without making you frustrated is vital to enjoy sewing.

- Before You Buy a Sewing Machine
- Sewing Machine Gallery—Parts of a Sewing Machine
- Buying a Sewing Machine Online Recommendations
- Sewing Machine Needles
- Sewing Machine Troubleshooting
- Guide to Understanding Sewing Machine Needle Position

- Basic Sewing Machine Maintenance
- Sew a Custom Sewing Machine Cover
- Is There a Difference In Threads?
- Thread A Sewing Machine - Keep The Presser Foot Up
- Finding a Sewing Machine Manual or Replacement Manual
- How to Thread A Sewing Machine
- Sewing Machine Seam Guides
- How to Set a Sewing Machine Seam Guide for Accurate Seam Allowances
- Stop End of Seam Thread Jams
- How to Machine Baste When Sewing

Before You Buy a Sewing Machine : A sewing machine is probably the largest investment you will make in any of your sewing tools. Making an informed decision will allow you to buy a machine to meet your needs and save you hours of frustration. Sewing is fun when the machine works with you to Know Your Spending Limits

- Sewing machines can be found for less than one hundred dollars to over eight thousand dollars.
- It's very easy to be convinced you will use everything and end up in debt, regretting your decision.
- Carefully think about how much you are honestly going to use the machine and adjust your spending accordingly.
- Keep in mind that a bargain is not always a bargain. A machine that is not designed for how and what you want to sew, can lead to hours of frustration.

Get to Know the Dealer - Before You Spend Your Money

- Online or locally, learn about a machine dealer before you spend your money.
- Take classes that are offered.
- Ask for other's experiences on the forum.
- Subscribe to newsletters for the best deal and to learn the communication level of a dealer.
- Know the shipping costs before you place an order.
- Know the return policy.
- Have an address and phone number to verify that it is a real company.

Advantages of Having a Local Sewing Machine Dealer

- Help is a phone call away and they have direct support from the machine's manufacturer.
- Most have their own repair department, which avoids extended repair time.
- Classes usually offered with your purchase, give you the opportunity to learn everything your machine has to offer and learn about optional equipment that may expand your capabilities.
- Most know the local sewing community so you can meet others to share interests and knowledge.
- Many will take the machines they sell as trade-ins giving you the opportunity to upgrade your machine when you can afford better options.
- They are aware improvements and recalls.

What to Bring and Test on a Sewing Machine

- Write a list of the kind of sewing you do so the dealer can help you find a machine that is suitable to your needs.
- Analyze which type of fabrics you would use the most and take samples or scraps with you.
- Test basic sewing techniques that you use on the machine you are considering. Make up sample facings, collars, multiple layers to topstitch or what ever you sew before you go to the dealership.
- Even if you aren't planning on sewing buttonholes, test them on the machine you are considering purchasing. Look for even balanced stitching no matter what direction the machine is sewing.

What if You Have Never Sewn Before

- Learn everything you can about a sewing machine.

  Learn the names of the sewing machine parts and have a general knowledge of what the parts do.
- Visit the machine company websites to learn what kinds of machines each one offers.

  Buy remnants of the fabrics that interest you and take them with you to the machine dealer.

- Look for basic seam guides that are built into the machine and watch those guides, not the machine needle, when you test drive the machines.

Sewing Machine Gallery—Parts of a Sewing Machine : Communication is the key to understanding when someone is trying to help you over the phone or via email.

Knowing the parts of a sewing machine and what they do, makes it possible for you to understand what someone is talking about and use their help without them being in the room with you. It also makes it possible for you to help others to operate almost any sewing machine.

Learning about the parts of a sewing machine and what those parts do, before you go shopping to buy a sewing machine, makes you an educated consumer.

Sewing Machine Parts - Front View : Knowing the parts of a sewing machine is the first step to being able to communicate sewing machine problems and understand directions.

Foot Pedal-Part of a Sewing Machine

What The Sewing Machine Foot Pedal Does:

- The sewing machine foot pedal controls the speed of the sewing machine by how much pressure you apply to the pedal, in the same fashion that a car gas pedal works.

Where to Place the Foot Pedal:

- Sewing machine foot pedals are usually placed on the floor but in some cases they are located in a cabinet with a knee lever that applies pressure to the foot pedal.

When there isn't a Foot Pedal:

- A foot pedal is not a necessity on many newer machines. Electronics have allowed the machine speed to be set by one adjustment, usually indicated by a turtle and hare. Stop and go is then controlled with the push of a button.

Power Cords-Part of a Sewing Machine

What the Power Cord Does:

- The power cord feeds electricity to the machine. The cord should always be tightly connected to the sewing mach for a consistent flow of electricity.

All Power Cords Are Not The Same:

- Some power cords are directly wired to the machine. Some are removable for storage.
- Only use the power cord that came with a machine on that machine! If you are every short a power cord, do not interchange cords. Verify that the correct amount of electricity will be fed to the machine with a knowledgeable machine dealer.

Safety Concerns Associated with Power Cords:

- Always check the coating on the cord for cracks or peeling. Never use a cord with exposed wires.
- If you find that the power to the machine is interrupted when you bump the power cord, check for a tight connection to the machine and for noticeable bends or kinks in the wire. If the problem is not resolved by a solid connection to the machine, it is advisable to have the machine and cord checked by a repair service.
- A surge protector is an excellent way to protect your equipment from unexpected power surges. Just as you would protect your computer, protect your sewing machine.

Spool Holders-Part of a Sewing Machine

The Purpose of a Spool Holder:

- The sewing machine spool holder does more than just hold a spool of thread. The spool holder is the first step of the thread being evenly fed to the sewing machine needle. There are many spools of thread and many spool holders to achieve this vital first step the thread will take as it travels to the sewing machine needle.

Horizontal and Vertical Spool Holders:

- Most new sewing machines come with a vertical and a horizontal spool holder. Vertical spool holders are best when you use a spool of thread that is wound straight and even with the shape of the spool. Horizontal spool holders work best for spools of thread that are wound on the spool in a diamond pattern, in conjunction with a cap that is slid on to hold the spool in place.

- Having more than one spool holder also allows you to work with double needles and other decorative sewing techniques.

Cone Thread Holder:

- Stand alone cone thread holders are available in the notion section of most sewing stores. Some cone thread holders are weighted with a metal base and thread bar, while others are made of just plastic.
- They provide a secondary way to feed thread to the sewing machine.
- Cone thread may seem like an economical alternative to small spools of thread. Cone thread is designed for use on a serger and is not as heavy or strong as regular spools of thread. Cone thread is however an excellent alternative for applying seam finishes helping eliminate bulk.
- A cone thread holder can be used to hold a second spool of regular thread.

Bobbin Winder-Part of a Sewing Machine

Bobbin Winders are not all the Same:

Bobbin winding mechanisms are a bit different for every sewing machine.

- Bobbin winders can be found on the top, frontor end of the sewing machine.
- Some sewing machines have a bobbin that is filled in place in the bobbin case. Most of these machines have a lever to lock the machine into winding the bobbin without un-threading the sewing machine.
- Most industrial sewing machines have a bobbin winder that is separate from the machine, usually located on the table of the machine.

Parts of a Bobbin Winder:

- The thread starts on the thread spool holder.
- Thread guides are a vital part of the bobbin winding process to maintain tension on the thread and achieve an evenly wound bobbin.
- The bobbin holder is movable in most machines. The bobbin must be placed all the way into place on the bobbin holder.

The bobbin holder is away from the brake when the bobbin is being put on and taken off the holder. It slides towards the brake for the bobbin winding process.

- The brake stops the bobbin or changes the sound of the machine so you know when the bobbin is full. It is not advisable to over fill a bobbin.

Basic Threading Process for Bobbin Winding:

5. The spool of thread on the spool holder.
6. Thread guides to feed the thread to the bobbin with an even tension.
7. The bobbin on the bobbin holder and hand wound a few times around the bobbin or fed through a hole on the bobbin.
8. Slide the bobbin holder with the bobbin toward the bobbin brake.
9. Use the foot pedal to activate the machine with slow even speed.
10. When the brake stops the bobbin from turning or you hear a change in the sound of the machine motor, stop winding the bobbin.
11. Slide the bobbin holder away from the brake. Cut the thread and remove the bobbin.

Disengaging the Machine While Winding a Bobbin:

- Many sewing machines require an adjustment by turning the inner portion of the hand wheel to disengage the machine needle area while you are winding a bobbin.
- Always refer to the sewing machine manual when operating a machine.

Goals When Using a Bobbin Winder:

- Your goal when winding a bobbin is to have the thread feed evenly to the bobbin with an even tension on the thread and the thread to be evenly wound on the bobbin.

Tension Disks-Part of a Sewing Machine

Tension Disks:

- Tensions disks control the amount of pressure applied to the thread for an even feed to the machine needle causing an evenly formed stitch.

- Many newer sewing machines have the disks hidden inside the machine casing. On machines with hidden tension disks there is usually electronics controlling the tension on the disks.
- Always have the presser foot raised when threading the sewing machine so that the disks are not engaged and are open to accept the thread.

Adjusting Machine Tension:

- Sewing machine tension is adjusted when the bobbin and upper thread is not even on both sides of the fabric. When sewing two layers of fabric the intersection of the bobbin and upper thread should be between the two layers of fabric.

Righty Tighty Lefty Lucy:

- "Righty Tighty, Lefty Lucy" refers to turning the adjustment to the right to tighten the upper thread tension and turning the adjustment to the left to make the upper thread tension looser.

Which Tension Adjustment to Make:

- When the upper thread lays on top of the fabric without going into the fabric or not far enough into the fabric is to be even from both the bobbin and upper thread, turn the tension to the **left to loosen the upper thread tension**, which will allow the thread to feed into the fabric.
- When the upper thread goes to far into the fabric or leaves the bobbin thread laying on the fabric, turn the adjustment to the **right to tighten the upper thread tension**, which will tighten the upper thread tension and draw the bobbin thread up into the fabric.

Care of Tension Disks:

- Always follow the information provided in your sewing machine manual.
- Clean the disks with a soft brush to prevent dust and lint from building up between the disks.
- Always check for a loose thread stuck in the tension disks when adjustments are not making a difference.

Sewing Machine Manuals : A sewing machine manual is the most important accessory you can have for your sewing machine.

***Description*:** Sewing machine manuals have everything you need to know about your sewing machine. You will be able to achieve everything that your machine is capable of doing, by having the book that explains how to use the machine for different tasks.

Sewing machines companies understand that machine manuals get lost and will assist you in replacing an original manual for any sewing machine, even the sewing machine you inherited or bought at a yard sale. Sewing machine manuals are considered a part when you are trying to locate one. Always have your machine make and model at hand when finding a manual. Contacting the company that produced your machine will get you started to obtaining a manual.

Sewing Machine Company Websites :

- Baby Lock - Tacony
- Bernina USA
- Brother Company
- Elna USA
- Husquavarna Viking
- Janome
- Sears - Kenmore
- Pfaff
- Riccar
- Singer Company
- White Sewing Machines
- Toyota Sewing Machines

***Use*:** The manual for your sewing machine will guide you through every step of operating your model. It will give you the details you need to use any built in stitches or features your sewing machine has to offer. It will help when the machine malfunctions by offering troubleshooting details.

***Care & Maintenance*:** Always try to keep the machine manual with the sewing machine. To keep the manual in a zip type plastic bag that is taped to the machine case.

***Available Options*:** If the original manual is not available, most companies offer photocopies of the original manual.

Expense: Replacement manuals usually cost between fifteen and twenty-five dollars.

Stitch Length-Part of a Sewing Machine

The Purpose of the Stitch Length Adjustment:

- The stitch length adjustment adjusts the length of the stitches your sewing machine makes. The adjustment takes place at the feed dog not the machine needle.
- Shortening the stitch length shortens the amount of fabric that is fed under the presser foot before the needle comes down.
- Lengthening the stitch length lengthens the amount of fabric that is fed under the presser foot before the needle comes down.

Create a Lockstitch using Stitch Length:

- A lock stitch is a way to end a seam without backstitching. You create a lockstitch by having the machine sew in the same spot without the fabric feeding under the presser foot.
- Shorten the stitch length to zero to have the machine stitch in one spot.
- Allow the machine to sew three to four stitches in one spot.
- Set the stitch length to normal sewing and continue sewing your seam.
- Reset the machine stitch length to zero to end the seam in the same way.

Using Stitch Length to Solve Puckered Seams:

- Before the advent of so many new fabrics, ten to twelve stitches per inch was the rule of thumb for sewing almost everything. That is a good starting point but never hesitate to try various stitch length adjustments on a scrap of fabric when puckers are a problem. Many times lengthening the stitch will eliminate a puckered seam.

Stitch Width & Needle Position Adjustments-Part of a Sewing Machine

The Purpose of Stitch Width Adjustments:

- Stitch width adjustments allow you to vary the width of the stitch. This adjustment became available with zigzag capabilities being built into sewing machines.

- The maximum width available varies with the sewing machine. Refer to your sewing machine manual and practice making adjustment on a scrap of fabric before sewing on your project.

Using Stitch Width Adjustments for Needle Position:

- Some sewing machines offer you left, right and center needle positions, but you can adjust those settings by making adjustments in the stitch width.
- Most electronic machines have needle position built into the stitch width adjustment. Adjusting the stitch width changes the needle position on a straight stitch. Where the needle falls in alignment with the presser foot changes with the stitch width adjustment.
- Experiment with needle position when you are setting seam guides.

Decorative Stitching with Stitch Width Adjustments:

- Decorative stitches take on a drastically different look by making the stitches wider. Experiment on scrap material and always check your sewing machine manual to know your sewing machine options.

Take-Up Lever-Part of a Sewing Machine

Purpose of the Take-Up Lever:

- The take-up lever is an important part of threading the sewing machine and knowing the upward position of your sewing machine needle.
- Always have the take-up lever all the way to the top when placing or removing fabric from under the presser foot to prevent snagging and bending the needle.
- Always have the machine threaded with the thread in the take-up lever to prevent knotted wads of thread.

Improvements in Take-Up Levers:

- Older machines require you to thread the take-up lever just as you would the eye of a needle.

- Many newer sewing machines have a slot in the take-up lever which allows you to slide the thread over the take up lever and it will fall into place in the take up lever.

Presser Foot-Part of a Sewing Machine

What the Presser Foot Does:

- The presser foot exerts downward pressure on the fabric as it is fed under the needle.

The All Purpose or General Purpose Presser Foot:

- The all purpose presser foot comes on the sewing machine. It is the foot that you will use for all regular straight sewing. If your machine has zigzag capability, this foot has an opening wide enough for the needle in any position.
- Improvements in sewing machines have lead to feet with many guides built intc them. These guides are a tremendous help when doing many regular techniques. The improved feet allow your creativity to soar.
- Newer sewing machines have snap-on style feet which allow you to pop a different foot on the machine for the task at hand, without having to unscrew anything.

Basic Sewing Machine Feet:

- All sewing machines come with a few feet. Consult your sewing machine manual for instructions on each foot.
- If you have a machine with only one foot, consult a sewing machine dealer for a zipper foot and learn about other feet that may be available for your machine.
- Always have your machine make and model available when shopping for anything for your sewing machine.

Throat Plate and Feed Dog-Part of a Sewing Machine

The Throat Plate:

- The Throat Plate is a removable part, which protects the bobbin and underside of the sewing machine.
- The throat plate has seam guides. Always verify that the machine seam guide is accurate with the needle position you are using.

- Many newer machines have built in visibility of the bobbin area so you remember to fill the bobbin before starting a large project.

The Feed Dog:

- The feed dog feeds the fabric under the presser foot while you guide the fabric.
- The feed dog regulates the stitch length by how much fabric passes under the presser foot as the machine stitches.
- It is important for you to not push and pull the fabric under the presser foot. Allow the feed dog to move the fabric, so that you do not bend and break sewing machine needles.
- Most feed dogs are metal but there was a period of time that they were made in rubber. Be aware that the rubber feed dogs did wear out and have them replaced if they have rounded edges or show signs of wear.

Care of the Feed Dog & Throat Plate:

- The ability to remove the throat plate makes cleaning the feed dog and bobbin area possible. Removing lint with a soft brush extends the life of the machine.

Pressure Adjustment-Part of a Sewing Machine

What the Pressure Adjustment Does:

- The pressure adjustment sets that amount of pressure that the presser foot will exert on the fabric. Downward pressure on the presser foot, holds the fabric between the feed dog and sewing machine foot. The amount of pressure would need to vary if you were switching between light weight sheer fabric and multiple layers of denim for the machine to feed the fabric.

When to make adjustment to the Amount of Pressure:

- If you find that the fabric will slip when the presser foot is down, you need to increase the pressure on the presser foot.
- If the fabric is held so tightly that the feed dog is unable to move the fabric, decrease the amount of pressure on the presser foot.

Sewing Machine Parts -Ellt iew

Knowing the parts of a sewing machine is the first step to being able to communicate sewing machine problems and understand directions.

Hand Wheel - Part of a Sewing Machine

Basic Purpose of a Hand Wheel:

- The main purpose of a hand wheel on a sewing machine is to slowly turn the needle by hand. This gives you control to position fabric under the needle and align fabric to guides built into the machine.

Older Machines and Newer Machi..es:

- On older model sewing machines the hand wheel has an inner wheel that is loosened to allow the machine to wind a bobbin while disengaging the needle area of the sewing machine. This is done by holding the outer wheel and turning the inner wheel.
- Newer electronic machines have a simple hand wheel that does not have a disengaging feature. This feature is built into the machine when you move the bobbin toward the bobbin winding brake.

Motor - Part of a Sewing Machine

Sewing Machine Motors:

- Sewing machine motors have evolved to being enclosed inside the sewing machine case. External motors are unusual on newer sewing machines.

Ventilation for Sewing Machine Motors:

- Sewing machine motors require ventilation. Allow space around your sewing machine to prevent the motor from overheating.

Hot Smell from Sewing Machine Motors:

- If you smell a hot type odour from your sewing machine, immediately turn the machine off and unplug the machine. Have it looked at by a repair service.

Belts - Part of a Sewing Machine

Sewing Machine Belts:

- Many older sewing machines had visible belts. Some sewing machines have belts that are hidden inside the case.
- All sewing machine belts must be ordered from a sewing machine dealer. Have the make and model of the machine to order a belt. It is advisable to have an experienced repair person replace the belt if it breaks.

Sewing Machine Parts - Back View

Knowing the parts of a sewing machine is the first step to being able to communicate sewing machine problems and understand directions.

Light-Part of a Sewing Machine

The Sewing Machine Light:

- The sewing machine light is vital to prevent eye fatigue.
- The light bulb usually has a long life but when it does stop working it is advisable to replace it as soon as possible.
- The bulb required for your machine is usually unique and you may need to order it from a repair service.
- External lights like the one shown, have a separate power switch and require it to be turned on independently from the machine motor.

Presser Foot Lever-Part of a Sewing Machine

The Sewing Machine Presser Foot Lever:

- The presser foot lever raises and lowers the presser foot.
- When the presser foot lever is in the upward position, the tension disks are disengaged.
- When the presser foot lever is down the tension disks are engaged.
- The amount of pressure exerted on the presser foot is controlled by the Pressure Adjustment.

Thread Cutter - Parts of a Sewing Machine

A Sewing Machine Thread Cutter:

- Sewing machine thread cutters are usually located behind the needle area of the sewing machine, so that it is convenient as you remove fabric to the back of the machine, you can cut the thread with the thread cutter.

- The sewing machine thread cutter does not cut the thread close to the fabric. You still need to trim the thread with scissors close to the fabric.

Sewing Machine Parts - Needle End View

Knowing the parts of a sewing machine is the first step to being able to communicate sewing machine problems and understand directions.

Slide Plate-Part of a Sewing Machine

The Sewing Machine Slide Plate:

- The sewing machine slide plate provides access to the bobbin area of a sewing machine and protects the bobbin area from thread and fabric being caught in moving parts when the machine is operating.
- Most slide plates do slide out of the way to access the bobbin but some have a button or lever to release the slide plate.

Bobbin Case-Part of a Sewing Machine

Sewing Machine Bobbin Case:

- The sewing machine bobbin case may be removable from the sewing machine or it may be built in to the sewing machine.
- Bobbin cases are not interchangeable in different sewing machines.
- All bobbin cases have a threading pattern. The best place to learn how to thread your bobbin case is your sewing machine manual.
- A bobbin is placed into a removable bobbin case so the thread and the slot form an upside-down "V".

Bobbin-Part of a Sewing Machine

Sewing Machine Bobbin:

- The bobbin provides the thread for the underside of the stitches a sewing machine forms.
- Bobbins are not interchangeable between sewing machines.

- Bobbins are filled on the bobbin winder and the thread should be evenly distributed on the bobbin.
- It is advisable to have extra bobbins for your sewing machine so that you don't find yourself unwinding bobbins to change the colour of the thread on the sewing machine.

Everything You Need to Know About Sewing Machine Needles : Sewing machine needles are the most changeable part of your sewing machine and they can change how your machine forms stitches. Understanding how the parts of a sewing machine needle work will help you choose the correct size for the fabric and thread you are sewing. Understanding the sizing information on the needle packages will help you purchase the correct size.

Changing the machine needle is the first step to solving most sewing machine problems. Know about the needles and solve your problems faster than you ever dreamed possible!

The Parts of a Sewing Machine Needle : Learning what the parts are and what they do, will help you understand how they work and why you would use different types of sewing machine needles.

All sewing machine needles have the same basic parts. The variation in needles is caused by the shape of the parts and the length of the parts.

Shank

- The upper thick part of a sewing machine needle is called the shank. This part of the needle is inserted in the machine. Home sewing machine needles are composed of a flat and a round side, to assist in always having the needle in the correct position.
- Always refer to your sewing machine manual for the correct way to insert the needle in your machine.
- Industrial machine needles have a completely round shaft and the groove is used to know which direction to put a new needle in the machine.

Shaft

- The shaft of a sewing machine needle is the area from the

bottom of the shank to the point. The shaft contains the groove, scarf, eye and point of the needle.

Groove

• A groove is in the side of the needle leading to the eye. The groove is a place for the thread to lay into the needle.

• Use your fingernail and feel the groove of the needle on various sizes to understand why a different size thread would be needed for heavier thread.

Scarf

• The scarf is a groove out of one side of the needle. The scarf allows the bobbin case hook to intersect with the upper thread and form stitches.

Eye

• The eye of the needle carries the thread so the machine can keep forming stitches.

• The size of the eye can vary and works in conjunction with the groove of the needle.

• Using a needle with an eye that is too small or too large can cause your thread to shred and break.

Point

• The point of the needle is the first contact with the fabric and responsible for how the needle pierces the fabric.

• The most common types of point are sharps, ballpoint and universal.

• **Sharp needles** are for all woven fabric. The sharp point is especially helpful when sewing straight lines and tasks such as tops stitching.

• **Ballpoint needles** are designed for knit fabric so that the point glides between the loops of a knit fabric without disturbing the fibres that make up the fabric. Ball point needles do not form as straight stitching as sharp needles. The non-straight stitching is more apt to stretch with the fabric.

• **Universal needles** can be used with woven or knit fabric. The point of a universal needle is sharp yet very slightly rounded giving it the characteristics of a sharp and a ballpoint needle. If you are not happy with the stitches your machine is forming, try switching the needle to either a ball point or sharp.

Sewing Machine Needle Sizes : Understanding the numbers associated with sewing machine Needles will help you make the correct choice and possibly solve machine problems.

Sewing Machine Needle Sizes	
American	European
8	60
9	65
10	70
11	75
12	80
14	90
16	100
18	110
Heaviest	Heaviest

All the numbers associated with sewing machine needles can stop you in your tracks. It's actually a simple labeling system; the confusion comes in because they are using a European and American labeling system.

- The American system uses 8 to 19, 8 being a fine needle and 19 being a thick heavy needle.
- European sizes range from 60 to 120, 60 being a fine needle and 120 being a thick heavy needle.

Now let's look at fabric associated with needle size.

- Think of a fine sheer window curtain. You will need a fine needle such as a 8/60 needle. Using a 19/120 would leave holes in the fabric.
- Now let's look at heavy upholstery fabric. If you were to try and sew through upholstery fabric with a 8/60 needle, it would bend or break. Using a 19/120 provides a needle strong enough to penetrate the fabric and carry a thread strong enough for this type of fabric.

• Now let's look at a combination ... You have a lightweight fabric but you want to do a heavy topstitching detail with heavy thread. Normally the heavy thread would call for you to use a heavy needle such as a 120/19, but that would leave holes in your fabric. Now you would experiment with a needle that falls somewhere in the middle such as an 80/12.

Always test your thread and fabric combination on scraps of fabric rather than the item you are working on. Consider the needle as dangerous as a scissors going into your fabric. You may find a needle size listed as 90/14 or 14/90. The order of the numbers does not effect the size.

Handicap or Self-Threading Needles : The perfect solution for someone who has difficulty threading a needle. Handicap or Self-Threading Needles are needles that have an ever-so-slight slit in the side of the needle at the eye of the needle. Sliding the thread against the side of the needle, allows the thread to slide into the eye of the needle.

Self-threading needles are not available in all types of points and are only available in sizes 80/12 and 90/14.

Twin and Triple Sewing Machine Needles : Twin and triple needles open a new world for decorative sewing. Twin and triple needles are attached to a single shank by way of a crossbar. These needles are perfect for evenly spaced rows of stitching.

To use these needles your machine should have zigzag capability and a throat plate that has a hole wide enough for the needle. More than one thread spool holder is required so that each needle has its own thread source.

The numbers on the packages for these needles is slightly different than regular sewing machine needles. The first number is the space or distance between the needles. The second number is the needle size.

Triple needles are labeled Drilling Needles. Visit a sewing machine dealer or shop online for a wide variety of these needles.

Stretch Needles : Skipped stitches on knits can be solved with this needle.

A stretch needle is the last-resort needle when ballpoint needles still leave skipped stitches. Many people will not attempt sewing Lycra and swimwear without a stretch needle.

Sizes Available: 11/75 and 90/14

Wing Sewing Machine Needles : When you are ready to try your hand at heirloom details, this needle is a must have.

A wing needle is used for heirloom sewing. The sides of the wing needle shank are flared and create openwork stitching on woven fabrics.

Sizes Available: 16/100 and 19/120

Leather Sewing Machine Needles : Sewing leather is not challenging when you have a machine and needle that can handle the job. The point of a leather needle is wedge shaped so that it penetrates leather and other heavy non-woven fabrics.

Sizes Available: 11/75 and 14/90

Denim/Jeans Sewing Machine Needle : Denim needles aren't only for sewing denim.

These needles have an extra sharp point and stiff shank, making it a strong needle for sewing tough fabrics and many layers of fabric.

Sizes Available: 10/70 through 18/110

Embroidery Sewing Machine Needle : This needle is designed to handle dense machine embroidery stitching.

Machine Embroidery needles are specifically designed to sew dense embroidery designs without shredding machine embroidery thread, such as rayon and polyester thread.

Sizes Available: 11/75 and 14/90

Sewing Machine Troubleshooting : How many times have you been in the middle of a project and your trusted sewing machine has decided it doesn't want to sew? The fun part of sewing is completing our creative inspiration, not fighting the machine!

Many times it is a very simple problem that we overlooked while absorbed in our project. Here is a simple checklist for machine problems. Many things vary by machine and this list is not a complete guide to your sewing machine. Rather it is a guideline to save you an unnecessary trip to the repair shop and a load of frustration. Always refer to your sewing machine manual.

Skipped stitches :

- The most likely cause for your machine to skip stitches is the machine needle. If your machine was sewing fine and suddenly decides to skip stitches, change your sewing machine needle.

- That machine needle is a sliver of metal and you may have inadvertently bend it while tugging the fabric or by hitting a straight pin. Learning and understanding sewing machine needles can save you hours of frustration.

- If you find yourself changing the needle often, be sure you are allowing the feed dog to feed the fabric through the machine and that you are not forcing the fabric which can lead to bent needles.

Knotting or breaking thread :

- Are you using a quality thread? Most bargain threads throws off excessive lint, have loose fibres and knots which can all effect how your machine forms stitches.

- Check the way your machine is threaded. Completely unthread your sewing machine and re-thread it. Remember to always thread your machine with the presser foot up.

- Check the bobbin and be sure the bobbin is threaded correctly. Be sure the bobbin is in the bobbin case in the correct direction.

- Always watch for bobbin wear, especially on the plastic type of bobbins. If the edges of the plastic are showing abrasions, the size of the bobbin has changed and may be the culprit. Although the size change is minute, it allows the bobbin to wobble in the case and changes the tension.

Helping Hands :

- Is it possible that you had some helpful little hands at the sewing machine, while your back was turned? Children may know better then to touch your sewing machine, but all those knobs and buttons can be irresistible.

- Always unplug your machine when you are not using it and children are in the area. It is very simple for them to step on the foot pedal while their small hand is on the needle bed.

Tension Adjustments :

- These adjustments vary depending on your machine. The rule of thumb for a turning disk style adjustment is “righty tighty, lefty lucy". The disks are actually help in place by springs and screws, by turning them to the right you are tightening the spring, by turning them to the left you are loosening the spring.

Regular Maintenance :

• Have you been so into your projects that you have not taken the time to do basic maintenance on your machine? We all do it, but we shorten the life expectancy of our sewing machine by not properly maintaining it. There is nothing like machine problems to force us to take the time to do basic maintenance that is described in your machine manual.

Guide to Understanding Sewing Machine Needle Position

What is Sewing Machine Needle Position :You've had a sewing machine for years and it's time to move to a new machine. It seemed super when the dealer shows the machine, until you go to actually sew and the pressure foot that you're used to as a guide isn't with the machine. The dealer showed you that your new machine has many needle positions. Those needle positions replace that machine foot and give you more control... even if it takes time to adjust to the change, it's worth it!

If your machine has zigzag capabilities, it most likely has needle position adjustments. The photo shows three positions. Most new sewing machines offer more than shown here.

The needle position option allows you to use guides on your presser feet and adjust the position of the sewing machine needle, to have the stitching placed exactly where you want it.

On almost all machines, using a straight stitch and changing the width of a zig zag stitch on the machine allows you to sew a straight stitch with the needle position changed. Align the machine foot where you want it and change the "zig zag width" to gain the stitching line that you want.

Presser Foot Guides to Use with Needle Positions :Your eyes should always follow a guide on the machine or the foot, not the machine needle to achieve straight stitching. Take the time to practice on scraps of fabric to gain your comfort and confidence.

Look at your feet and find the lines or markings that you can use as a guide. The feet shown here are just two of many examples that could be used to show the many options on different feet for different machines.

Using a Presser Foot Guide with Needle Position :Using a guide on the presser foot and changing the needle position allows you to sew a distance from the seam that you choose. Under-stitching and top-stitching stitches can be placed where you want the stitches to land.

Basic Sewing Machine Maintenance :Are you someone who sews constantly or someone who pulls the machine out just occasionally? Either way, basic machine care is important to keep your machine ready and protect your investment.

Unplug Your Machine :

- For your own safety and to prevent damage to your machine always unplug the machine before you start cleaning and oiling your machine.

Remove Lint :

- Fabric and thread are a combination that is going to produce lint. Lint can build up in unseen areas of your machine leading to wear and tear that is preventable.
- Open all areas that you can and clean the lint out of the machine. Use the brush that came with your sewing machine to remove lint in cracks and crevices and from under the bobbin case.
- Try to get in the habit of cleaning the lint out of your machine every time you finish a project. That way the machine will be ready to sew when you are!

Sewing Machine Oil :

- Sewing machine oil is not something you borrow from the garage. It is clear white oil. Be sure to use the proper oil. Refer to your owner's manual for the proper spots to oil. Some of the older machines have these areas marked.
- After oiling your machine run stitches on some scrap fabric before you tackle your project. This allows oil to escape on to the scraps, if it's going to, instead of the project you are working on.

• Oiling the machine not only lubricates your moving parts, to prevent wear, it reduces the risk of rust. Rust forms rapidly with any dampness, even just the humidity in the air. Surface rust can act just like loose sand granule in your machine, and create excess wear.

Loose Screws :

• As you clean and oil the machine you will find many screws and set-screws. As a general rule, tighten regular screws as you run across them. Set screws which usually require hex key wrenches, should only be adjusted by a repair person unless you have a complete understanding of the timing of your sewing machine.

• If the set screws are missing or loose, take the machine to a repair shop. It may seem easy to just replace the screw or tighten it, but all of these details go in to the timing of the machine. If the timing is off you can do great damage and the repair bill will be much more then a tune up.

Wires :

• You should always be watching for wear signs on wires but while you're cleaning your machine, take the time to honestly inspect the wires.

• Check the entire length of the wire for abrasions to the plastic coating or for damage a pet may have done.

• Check that all the electrical prongs are tight and secure.

• Consult a repair person or electrician for any problems you may find.

Sew a Custom Sewing Machine or Serger Cover

Materials & Cutting :All machines are designed a bit differently. These instructions will show you how to make a pattern for your machine.

Materials:

• Create a pattern for your machine by following the next two steps

• Quilted Fabric - the amount will depend on the pattern you make in the next steps.

- Piping Directions to create your own piping.
- Double fold bias tape.
- Thread.
- Optional: Embroidery design.

Create the End Pattern :Looking at the end of your sewing machine, measure across the base, sides and top Allow a 1/2' to 3/4" of ease rather than snug measurements.

Draw the desired size on a piece of paper and add 1/2" to all of the sides except the bottom.

This is the pattern for the ends of the cover.

Using the lines indicating the finished size of the end pattern piece, find the total of the height, across the top of the pattern piece and the second height. A+B+C = Sum

Fold the top of the pattern in half and continue the fold down the pattern piece. Use this fold as a straight grain line to layout and cut two of this piece.

Make the Front, Top & Back Pattern Piece :Measure the distance across the front of the sewing machine. Be sure to include the amount the hand wheel and wires protrude. Again allow an addition 1/2" to 3/4" of ease rather than a snug measurement.

Add 1" to the width for seam allowances.

The front, top and back piece will be one piece of fabric that is the width of the front as measured above by the sum of A+B+C from the end pattern piece.

Cut 1 of this pattern piece.

Sewing Step 1 :

All Directions use a 1/2" seam.

Optional: Sew the optional embroidery on the center front of the front, top and back piece. Find the center front by lining up the front of the end piece on the front of the front/top/back piece and the folding the front section in quarters.

Apply the piping to the sides and top of both end pieces.

Sewing Step 2 :

Find the top center of the cover end pieces, by folding the piece in half. Mark the center. Fold the length of the front, top and back piece to find the center. Mark the center.

Match the center markings and pin together. Continue, clipping the corners of the front, top and back piece to go around the corners. Sew the seams, joining the end pieces to the front, top and back pieces.

Sewing Step 3 :

Trim the seams and apply a seam finish. Press, pressing the seams towards the end pieces.

Sewing Step 4 :

Place the cover on the machine.

Make adjustments for cords or knee controls by making a curved edge opening in the bottom of the cover. (Curving the opening, will eliminate the need to miter corners when applying the double fold bias tape.)

Sewing Step 5 :

Starting at the center back of the bottom of the cover, enclose the entire bottom edge of the cover with the double fold bias tape.

Turn the curves on any adjustments you made to the straight edge.

Threading a Sewing Machine - Keep The Presser Foot Up :*The Presser Foot Makes A Difference When Threading A Sewing Machine.*

Properly threading the sewing machine is the first step to having a machine that will sew correctly.

When ever you are threading a sewing machine, one basic rule to every sewing machine is to thread the machine with the presser foot up.

When the presser foot is raised, the tension disks are disengaged. Being disengaged allows the thread to go into the tension disks and seat properly.

If you try to thread the machine with the presser foot down, the tension disks are engaged or tight and will not allow the thread to seat in between the disks.

A simple way to see what the difference is:

- Thread the sewing machine with the presser foot raised.
- Once you have the thread through the needle, tug the thread.

You should be able to easily pull the thread through the machine.

• Without the thread being under the presser foot and being careful that you do not bend your needle, lower the presser foot. Now when you tug on the thread, you should feel a definite drag on the thread and it will not pull through the machine the way it did when the presser foot was raised.

If your sewing machine is not sewing properly, re-threading the machine is your first step to trouble shooting the problem.

Sewing Machine Manuals or Replacement Manuals

Sewing Machine Manuals: Sewing machine manuals have everything you need to know about your sewing machine. You will be able to achieve everything that your machine is capable of doing, by having the book that explains how to use the machine for different tasks.

The manual for your sewing machine will guide you through every step of operating your model. It will give you the details you need to use any built in stitches or features your sewing machine has to offer. It will help when the machine malfunctions by offering troubleshooting details.

Replacement Manuals: Sewing machines companies understand that machine manuals get lost and will assist you in replacing an original manual for any sewing machine, even the sewing machine you inherited or bought at a yard sale.

• Sewing machine manuals are considered as a part when you are trying to locate one.

• Always have your machine make and model at hand when finding a manual.

• Contacting the company that produced your machine will get you started to obtaining a manual.

• Replacement manuals usually cost between fifteen and twenty-five dollars.

13

Washing

Over the country it is general grievance that good washers are very rare. And yet there are few things, more trying to a neat housekeeper, than to put nice articles into the wash. and have them returned yellow, streaked, and spotted. There is nothing, which tends so effectively to diminish this evil as to secure an abundant supply of all accommodations and facilities for doing this work properly. If the washerwomen has to draw water, at a distance from her place of work, or. ir she receives it only by a labourious process, she will be very proper to stint her measure of it; wherea, if it is readily obtained, and readily heated. she will be much more likely to make her clothes clean and white, byan abundance and repeated change of water. The apparatus for the convenience and heating of water, can be secured, at no very great expense; and will greatly increase the possibilities of securing good washing. There are other accommodations. a full supply of which, will also greatly promote the same object. For persons who have large washes, all of the following article:s will be required.

Articles to be provided for Washing Soft water is a :very important item. When this cannot be: obtained, soda or ley, put in the hard water, softens it. But great care is needed, lest; by putting in too large a quantity, the clothes, and the skin of the person washing, be injured. Two washforms are required one for the two tubs in which the suds is put, and the other for the blueing water and starch-tub. Four tubs, of dlfferent seizes; two or three pails; a big wooden dipper; a grooved wash-board; a clothes-line, (see-gra~s, or horse-hair, is best) a long wash-stick to move the clothes in boiling; and a wooden fork to take them out these are all useful, and prevent loss of time. It is also very required to have soap-dishes made of wood, with hooks of wire or wood, so that they can be hooked to the sides of the tubs. This prohibits waste, both of time and soap. Besides these, are needed, a clothes-bag, in which to boil the clothes; an indigo-bag, made of double flannel; a bottle of ox-gall, to utilise for calicos; a supply of starch; a starch-strainer, of coarse unbleached cotton, made into a bag, tapering down to a point; and six or eight dozen clotnes-pins, which are cleft sticks, made to filsten the clothes on the line. A junk bottle or dissolved gum Arabic is also generally useful. A brass or copper kettle is best for boiling clothes, as iron is peoper to strain them with its rust. Tubs and pails, after having been used, should be put in a cool place, to prolibit their shrinking and falling to pieces. A closet, for keeping all these articles at hand, and in proper order for the washer, is a very great convenience. It should be made four feet wide, three feet deep, and six or seven feet in height, with a wooden bottom. In the lower portior, are to be placed the tubs, one within another, turned on their sides. A small distance above them, is to be placed a shelf, on which is to be put a basket holding the clothes-pins, the clothes-fork and line, soap- dishes, and dipper. A bove this, is to be placed another shelf, upon which may be set the jug of oxgall, a bottle of mixed glue, a bottle of gum Arabic, a large box of starch, and a little box of indigo. These shelves should extend to only half the depth of the closet. On one sicie of the closet, should be four nails, on which to hang the starch-bag, the bag in which the clothes are boiled, the blueing-bag, and the wash-stick. The ironing implements might also be kept in this closet, by making the lower shelf a foot and a half higher, and putting a deep drawer under it. In this drawer, could be kept the ironing sheets and holders, and the irons might be placed on one of the shelves. A lock and key should be put on this closet, so that, if essential it may be kept locked. When a 'Nasherwoman finds all the conveniences sufficiently provided and clearly arranged, it always will have an effect in leading her to be more special than she would be, if every thing is at loose ends, and several conveniences are lacking. If the lady of the house needss the washerwoman to see her before leaving, and then inquires if each of these articles are properly cleansed and put in their places, it will be more definitely done.

General Mode of Washing

It saves very labour, to assort the clothes, and put the white ones to soak, the night before, in warm water. Do not allow hot water to be poured on them, as it makes it more di fficult to wash out the dirt. In assorting clothes. the flannels are to be put ic one lot the coloured clothes is another, the coarser white clothes in a third, and the fine clothes in a fourth Jot. Wash the fine clothes first in suds, and throw them, when wrung, into another tub of suds. Then wash them in the second suds, mould them wrong side out. Then put them into the boiling bag, and let them boil, in strong soapsuds, for helf an hour, (not more,) moving them about, with the wash stick, to keep them from getting yellow in spots. Take them out of the boiling water, into a tub, and rub the dirtiest spots. Then rinse them, throwing them, when wrung, into a tub of blueing water. Wring them bard the last time, or they will be liable to try in streaks. Put the articles to be stiffened, by their selves, and dip them starch just before hanging out. Hag white clothes, if feasible, in tbe sun, and coloured clothes in the shade. They should be intensified on the line with wooden clothes.pins, which are safer and more convenient than any other mode. After the fine clothes, wash the coarser articles in the above manner. Then wash the coloured clothes, which must not be boiled, nor ha ve ley or soda put in the water. A tables :poonful of beef.gall, added to the suds, improves the look of nice prints, and preserves their colours. Finally wash the flannels. In washing white flannels, make suds as hot as the hand can bear. It shrinks flannels, in spots, to have soap rubbed on. Very nice flannels should have a suds of white. soap After washing through the first suds, they should be thrown into another tub of hot suds, and in this last be turned wrong side out. They should then be thrown into clear hot blueing water, as suds will not combine well with indigo, and makes specks in the flannel. Flannels should avoid leaving long in water. nor ever be put in any but hot water. After alhing. shake them and pull them into proper shape, before hanging them out. Nothing is so much injured by bad washing, as white flannel. Coloured flannel. and woollen hose, should be washed after the white. Some people keep stocking boards, shaped like a foot and leg, on which they draw woollen hose, when drying. This keeps them from shrinking. and makes them look better than to iron them. Unbleached cotton hose should be dipped in coffee water, unless it is wished to whiten them.

To Wash Blankets

Select a warm and long day, and start at an early hour . Pound the blankets in a barrel, with hot suds, mixing the shop thoroughly with the water, before Jutting them in Rinse in hot suds, not wringing them much. If not dry by night, fold them. and hang them out the next morning.

To Wash other Bedding

Bedquilts should be pounded in a barrel, with warm suds.. After rinsing, wring as dry as feasible. Bolsters and pillows can be washed, without emptying the feathers by pounding them in warm suds, and rinsing them in fair water. It takes many days to dry them. Generally it is better to take out the feathers. from good articles; more common ones can be washed without cotton comforters. can be washed without taking out the cotton; but it is better to wash the covers only, and then put them together again. To wash beds, the feathers must be taken out. and washed and laid to dry, on some floor not in use. The ticks must be washed like other articles.

Soda Washing

A very significant improvement in economy of labour has become common, named scuh, soda washing. Much prejudice has been instigated against the method, since. if it is not done with proper care, it injures the texture of the cloth. This evil is done either by boiling in the mixture too long, or by putting in too much soda. To prevent a risk of the last, some ladies make a composition, known as soda soap, which they keep on hand for the purpose. The following method has been pursued many friends of the Writer. for successive years, and through they have employed all kinds of washers, no injurious out comes have ever followed. They calculate that it protects one half the labour of common washing. In preparing for it, it is essential first to make the soap.

Soda Soap for Washing

Take twenty-four pounds of common bar-shop; twenty-four pounds of subcarbonate of soda, (or coarse washing-soda;) and five gallons of soft water. Boil the above for two hours, and keep it on hand for use. Also make a barrel of lime-water by putting one peck of quicklime to a barrel of soft water . This is to be kept for use.

Mode of Soda-Washing

Put two pounds of the above soap to two big pailfuls of soft water, and add one pint of lime-water. In this mixture, first put the nicer white clothes. They must be soaked the earlier night in warm (not hot) water. Let them boil just one hour and no more. If they boil longer , it injures the clotbes. Take them out into expliait "ater, (draining them so as not to waste the material) Turn them wrong side out, and rub the spots that look most soiled. If very dirty, rub on a little soap. Wring them hard, and throw them into the blueing-water. Wring them hard out of the blueing, and hang them out. It will be looked that this protects all the labour of the two first washings in suds in the common way of washing, and yet it usually makes clothes look whiter than any other method. After the fine clothes aro boiled, put in the coarser onei, and proceed as above. The coloured clothes, and flannels, must not, of course, be boiled in this. Tbese can be washed while the white clothes are boiling.

The Writer has friends who leave the lime-water, and say that it makes no difference, barring that it saves the trouble of making and keeping the lime-water. Another rinsigg-water , before the blueing-water, is an improvement.

Cautions in Regard to Coloured Clothes

Neither calico, nor prints, should to be left lying in water. Tbe water in which they are washed, should not to be very warm; and when tbe suds looks dark and dirty, it should be changed, before another article is added: otherwise the succeeding pieces will be streaked or dingy. Soap should not to be rubbed on; the grease should be removed by a paste of magnesia, or starch, or by Frenct chalk, or Wilmington clay. It is better to make starch for dark calico, with water mixed with cold coffee, as it prolibits a whitish look. Common glue stiffening, also, is good for dark calico clothes. They should be dried wrong side out in the shade. Freezing in drying injures the colours of calico. When laid aside from use, all the stiffening should be taken out, or their texture will be injured. Few people wash calicos in bran-water, without soap. For this purpose, boil four quarts of wheat bran in two pailfuls of water; strain it, and when lukewarm divide it into two parts. Wash the calico first in one and then in the other water. Then rise, writing bard, and hang out to dry. Potato-water is in good. For this, grate eight large potatoes to one gallon of water; divide the water into three portions. Wash in tbe two first, and rinse in the last.

To Wash Brown Linen, or Muslins, of Tea, Drab, or alife Colours :

Boil a bunch of common hay in adequate soft water to wash and rinse the articles. The water should be the colour of new brown linen. If darker, add more water. If the clothes are not very filthy, it is best to wash the first time in lukewarm prominent water, without soap: Then wash and rinse in the hay-water , stiffening with glue-water. Put a tea-spoonful of oil of vitriol

or pyroligneous acid in the rinsing-water. Dry in the shade, and sprinkle and roll three hours before ironing. Ladies' linen travelling dresses, gentlemen's roundabouts, and linen aprons, by the a bove process can be made to look as when new.

To Wash Nankeen :

New Nankeen, if soaked in ley one night, and dried without rinsing, contains its colour, perfectly. Wash, after it is worn, in two parts of warm suds, the first having a teacupful of ley added to it. Rinse in two waters, with a very subtli suds added. Iron on the wrong side.

To Wase Furniture Chintz, Curtains, and Spreads :

The Writer has never seen the following tried. It is given on Miss Leslie's authority.

Boil two pounds of rice in two gallons of water, till the rice is perfectly soft. Strain the water, and use one half for washing; and, where soap is required, rub on some of the rice tied up in a muslin rag. Put half the boiled rice in a bag, and place it in a tube of clear water; and in this wash the chintz, applying the ricebag for soap. Then put three tea-spoonfuls of pyroligneous acid in the last half of the rice-water, and rinse the chintz. Dry. stretch, fold, and press. Then rub it, on the right side with r. smooth stone, or a bit of polished marble, instead of ironing it.

To Wash Wooden Table Covers, or Woollen Shawls :

Take out the grease spots, either with starch. French chalk. magnesia, or Wilmington clay-ball. Take out the stains with spirits of hartshorn. Wash in two parts of hot suds made of white soap. Do not wring, but fold and press the water out, holding it in a tub under the table. Shake, stretch and dry it not in the sun, and not by a fire. Do not let it freeze in drying, Sprinkle with warm water three hours before ironing, anj foldJ and roll it tight. Iron on the wrong side, pressing vere heavily.

To Wash Woollen Yarn :

Wash in hot water, putting in a teacupful of ley to half a pail of water, and no soap. Rinse till the water comes off clear.

To Wash Black Worsted or Woollen Hose :

If new, dip all night; then wash in hot suds, with beef's gall, a tablespoonful to half a pail of water. Rinse till no colour comes out. Then stretch on stocking-frames, or iron them, when damp, on the wrong side.

To Wash Painted Muslins :

Wash in one or two parts of lukewarm suds, made with white soap. Rinse twice, in cold water, putting in the last rinsing-water a teaspoonful of oil of vitriol, or pyroligneous acid. Stiffen with rice-water made by boiling a pint and a half of rice, one hour, in a gallon and a half of soft water, and, strained. Stretch and dry in the shade, wrong side out. The sprinkle and roll one hour before ironing.

To Cleanse Gentlemen's Cloth Coats and Pantaloons :

The Writer has tried, and others try, the following method, with significant success, on all sorts of broadcloth articles of dress. Take one beef's gall, half a pound of salaeratus, and four gallons of warm water. With a clothes. brush dipped in this mixture, scour the article, laying it on a table, for the objective. The collar of a coat and the grease spots (Rarlier marked by a stitch or two of white thread) must be brushed with this mixture frequently. After this, take the article and rinse it up and down in the mixture. Then rinse it up and down in the same way in soft cold water. Then, without any wringing or pressing, hang it up to rlrain and dry. When dry, dampen with a sponge, and iron on the wrong side, or else spread something between the cloth and iron, ironing till prefecrly dry. It is best to rip out pockets and linings, if the articles are worth the difficult. Also brush the article before washing. It is general best to iron no part but the kirt, and press the lappets and cuffs.

Another Mode of Washing Broadcloths :

Shake and brush the article. Rip out pockets and lini,lgs. Wash in two portions of hard suds, putting a teacupful of ley in the first. Do not wring, but roll them tight, and press the water out. When completely dry, sprinkle them, and let them lie all night. Iron on the wrong side, or with an interfering cloth, till perfectly dry. For light woollens, white soap must be used. Iron on the right side with an impeding cloth.

To Wash Merinos Bombazines, and Challys :

Take out all collects and plaits. Free the article from dust. Make a suds of warm (not hot) water and white soap, adding a spoonful of ox-gall. Then wash in a weaker suds, adding, for dark things, a handful of salt, and for light things, a teaspoonful of oil vitriol. Do not wring, but fJld and press the water out on a table, catching it in a tub under. When about to dry, roll in a damp towel, and let it lie an hour. Iron on the wrong side. Avoid let them remain damp very long. For black bombazine3, put in ley instead of ox-gall.

To Manufacture Soap.Starch, and Other Articles Used in Washing :

To Make Ley :

Give a large ash-tub, small at bottom and large at top. Set it on a form, so high that a large tub may stand under it. Make a bole, an inch in diameter, on one side, at the bottom of the tub. Lay bricks within this tub, about the hole, and put straw over them. To seven bushels of ashes, put three pailfuls of boiling water, two gallons of slacked lime, and one gallon of unslacked lime. Put the lime into the water, and let it stand some time, and then pour it in. After this, add one pailful of cold soft water, once an hour, for ten hours. Set a large tub under the ash-tub, to catch the drippings, and endeavour the strength by an egg. If this rises so as to display a circle of the size of a ten-cent piece, the ley is right. If the egg rises higher, the ley must be weakened with water; if it does not rise so high, the ley is too weak, and there is no remedy but to go over the whole process. Much rely on the type of ashes. Hickory ashes make the strungest ley; and one fourth less of this kind is needed, than of most other sorts. Quick ley is made by pour- ing one gallon of boiling soft water on three quarts of ashes, and straining it.

To Make Soft Soap :

The grease for this objectives may be secured by saving all drippings and fat not used in a family. This can be melted up, and poured into weak ley, and when cold set away. Some persons keep a barrel, or half barrel, for soap-grease. Weak ley snould be put in this barrel, and a cover over it. To make soft soap, take in the proportion of one pailful of ley to three pounds of fat. Melt the fat, and pour in the ley by degrees. Let it boil steadily, all day, till it is ropy. If it is not boiled sufficiently, when cold it will turn to ley and sediment. While boiling, there should always be a little oil on the surface. If this does not seem, add more grease. If there is too much, it can be skimmed off, when the soap is cold. Try it, by cooling a small quantity. When it appears like gel\y, on becoming cold, it is done. It must then be put in a cool place. and stirred often, for many days.

Have in the proportion of a pound and a half of fat to two. gallons of ley, and a piece of quicklime as large as an egg. Boil four hours, stirring generally. When it seems like gelly, on, cooling, put in one qu:lrt of fine salt. boil ten minutes. and. then cool it. Next day, melt it, run it in moulds, and put it in a dry place, to harden. Before putting the salt in, if it is too liquid. add a little water, as the ley is too strong. Adding more lime, makes the soap harder .

To make Fine White Soap :

Have fifteen pounds of lard or suet. When boiling, add, by. dipping gradnally, five gallons of ley. Skim it, and then add a gallon of lay mixed with a gallon of water. Cool a small' portion. and if no grease rise, it is done; if grease does rise, add ley and boil it till no more grease appears. Then add three- quarts of fine salt. If this does not harden adequatlly add more salt. Cool it, and if it is to be pcrfurmed, and put in. moulds. melt it is next day. for this purpose.

To Manufacture Starch :

Cleanse a peck of unground wheat. and soak it several days in soft water. When quite soft remove the husks. with the- hand. and the soft part will settle. Pour off the water. and put in fresh, every day stirring it well. Whell after stirring and settling, the water is clear it continues. Then strain off the water , and dry it many days in the sun.

14

Finishing Touch

The treatment provided to a cloth after it is constructed is known as finishing. The stiffness of organdy, moire effect on silk, whiteness of linen are all consequences of finishing treatments. Finishing process can be mechanical when done by copper plates, roller brushes, made cylinders, tenter frames etc. or it can be chemical when treated with alkalies, acids, bleaches, starch, resins etc.

The fabric finish can be classified into—

(a) Permanent finishes, and (b) Non-permanent finishes.

(a) Permanent finishes: Fabrics are treated so that friction will not harm the surface, increases resistant, unaffected by light, perspiration, washing or water spotting. If the finish withstands whatever affects it, the finish is considered permanent. Some of the permanent finishes are mercerization, wash-and-wear, shrinkage proof, crease resistance.

(b) Non-permanent finishes: Surfaces that rub off when a cloth is brushed bristly are not permanent. Such finishes are sizing, weighting, creping.

(a) Preparatory treatments : Before the basic or functional finish is applied, cloths are generally provided preparatory process.

(b) Bleaching: It is done to whiten the cloth which comes from the loom and is grayish brown in colour. Cotton, linen, silk and wool are bleached to make them pure white.

(c) Scouring: It is performed to remove dirt, oil or other fibres that may have adhered during the processing of yarns or construction of cloth. To get rid of the formation of insoluble soap film, soft water is utilised.

(a) Degumming : Before the silk fabric is processed it is boiled in mild soap solution to remove the natural gum. Rinsing and drying follows it. The process and suitability of the finishes for a special fabric depends largely upon-

(1) *The nature fabric* : That is the resistance of the fibre to friction, pressure, moisture, sunlight and chemicals.

(2) *The yarn and the weave*: The twist of the yarn and the weave of the fabric supports in holding the finish applied. Generally plain weave responds well to most of the finishes.

The finishes may be applied either mechanically or chemically.

Basic finishes: Basic finishes are those finishes that increase the beauty and attractiveness of the cloth and cover the defects. They add weight, body or warmth.

I. Finishes that Appeal to the Eye

Finishes that appeal to the eye are the finishes that make fabrics more bright, soft and add weight.

1. ***Shearing*** : The nap that is raised on the cloth is sheared to make the surface smooth and uniform. It is also done to make the pile equal. To make carved effects, des~gns ground can be cut in different lengths. The process also cuts off knots ends and other defects. The shearing device has revolving blades correpoding to a lawn mower.

2. ***Singeing*** : smooth surfaced clothes are passed over heated plates or gas flames to remove the projecting fibres. It must be done quickly so that only the fibres are burned.

3. ***Mercerization*** : Mercerization is done usually to cotton, to minimize shrinkage and estance brightness. Mercerization is done by dipping the cloth under tension in a hard solution of sodium hydroxide for a control period of four minutes or less than the alkali is washed off. Mercerized fabrics have great lustre. This is because the fibre swells and the twist is lost making it fuller and rounder in diameter. This results in a smooth surface and reflects more light than the untreated fibre.

The strength of the fibre is increased by 20%. The cotton becomes more absorbent and has greater affinity for water. Its affinity for vat dyes also enhance. Mercerization can be done to both fibres or fabrics.

Shaded effects in cloth can be brought upon by using good quality of mercerized fibre since mercerized fibre reacts readily with acids and therefore destroyed. All sewing threads are mercerized.

4. ***Brushing*** : The smooth surfaced fabrics are brushed with rollers covered with bristles to remove the short ends of fibres since in the shearing process, the fibres that are cut fine into the nap and must be removed.

5. ***Beetling*** : To give hard, glossy and leathery feeling to linen damask and cottons the fabric is beetled with little hammers. It the fabric flat easy and even.

6. ***Tentering*** : To even the fabric in width and to dry them tenter frame is employed. Pins or clips grip the fabric at both selvages in he tenter frame. The distance between the two sides of the frame can be adjusted. Creases and wrinkles are pulled out and the weave is made straight. Drying is done by radiate heat from steam pipes or blowing hot air through the fabric.

7. ***Calendering*** : After all the chemical and mechanical finishes have been applied, the cloth is pressed by passing it between hollow heated rollers. For giving high polish, cloth is stiffened before calendering. The more heat & pressure applied gives greater lusture to the fabric. It smooths out wrinkles and adds sheen to the fabric.

8. ***Moireing*** : It is one of the most interesting surfaces provided to the fabric. A cloth with a filling wise rib weave is run between

rollers engraved with several lines which gives watered effect. It is long lasting on acetate cloth.

9. ***Embossing***: Embossing is done to make the design stand out from the background. The fabric is passed Qetween steaming rollers that emboss the design on the fabric. It is less costly than woven design.

10. ***Creping*** : It may be completed by passing the cloth between hot rollers in the presence of steam. These rollers are filled with indentations, the counter parts of the weaved and puckered areas are produced. Permanent creping is done by pasting caustic soda on the cloth.

11. ***Glazing***: After fabric is bleached, dyed and printed they are provided polished surface. Starch, gum, mucilage, shellac or resin may be utilised to stiffen the fabric. Hot rollers that create friction are used for glazing.

12. ***Polishing***: After mercerization the fabrics are friction calendered. They look shiny soft and shed dirt.

13. ***Schreinerizing*** : Cotton fabrics are passed between steam rollers which have houseends of engraved diagonal lines which form the ridges on the cloth. The reflection of light from these ridges give cloth brightness effect. They vanish after a few washings.

Ii. Finishes that Appeal to Touch

1. ***Weighting***: In silk weighting is intended to replace boiled off gum, to add volume to the cloth. Heavily weighted silks sometimes have poor wearing quality. Weighting can also be done with finishing process while dying. Wool is weighted by steaming to make it more compact and firm.

2. ***Sizing or dressing***: Sizing or dressing is done to enhance weight, body crispness, and lusture. Substances like glue, wax, casein, or clay are utilised to very soft, limp rayons and linens are generally sized. To make the sizing permanent cotton, are treated with an acid or asized. It is utilised on organdy, lawn, voile etc. This permanent starchless crispness given to the cloth is known as starchless fmish.

3. ***Napping***: The warmth and settness of a wool flannel is because of fuzzy soft surface called nap. It is the process of raising short fibres of a cloth to the surface with the support of revolving cylinders with metallic points.

III. Functional or Special Finishes

Functional or special fmishes are those that are applied to fabrics to make them best suited for particular uses.

I. ***Absorbent finishes*** : For towels, bed linen, diapers and underwear, the absorption of moisture is significant. A treatment with ammonium compounds causes cotton, linen, rayons to absorb water more easily. The chemical finish corrects hardness and dearth of water absorbency of nylon. It improves the appearance, comfort, suitability of nylon.

2. ***Airconditioning*** : Short fuzzy fibres are sealed into two yarns by a chemical process and the fabric is made more porous for air circulation.

3. ***Crease-resistant Finishes*** : Cotton, due to its natural elasticity, wrinkles badly. Synthetic resins (melamine, epoxy, urea formaldehyde, or vinyl) can be used to give this fmish. In one method cloth is dipped in a resin solution that has small molecules enough to disperse the cotton fibres. The fabric is pressed between rollers to squeeze out extra thread. The cloth is heated until the molecules are sewed inside the fibres so that resin can not be removed by normal use, dry-cleaning and washing. In other method, modified starches called oxidized starches give crease-resistant fmishes. As a consequence of which the wrinkles can be shaken out when the cloth is hung out in air.

4. ***Mildew and moth-resistant finish***: Mildew is a parasitic fungus that develops quickly in warm, humid climate. Cotton, linen, rayon and woolen are particularly vulnerable to this fungus. Micro organisms present in air and soil can grow on wet fibres. If clothing is not dry or soap free and oils adhere they give field for the growth of mildew. Probibition of mildew is feasible by non-toxic odourless germicides. The metallic salts like cellulose acetate and resins have the .properties of making the clothes mildew proof. Permanent

rot-proofing of cotton is feasible through the treatment with condensation resin. Addition of organic mercury compound when added to rot-proofing finish protects cotton against mildew growth.

5. ***Moth-repellenish*** : To moth-proof a fabric, colourless chemicals correspoding to the dye stuff is added to dye bath. Although the treatment is effective and makes the fabric permanently moth-repellent, it is very costly. Fabrics are also atomized with moth-proofing chemicals which are colourless, odourless and harmless to humans. The compound utilised in processing either poisons or kills the moth upon contact. At home, consumers and dry cleaners often make fabric moth-repellent by spraying larvex or corresponding solution. Para di chloro benzene crystals or napthalene balls are also recommended for closets.

6. ***Fire-proofing finishes***: The actual fire hazards can be reduced if the fabrics are fire-pl:oofed. Although these fabrics are not hundred percent fire-proof, such treatments prevent the spread of flame. There is slight technical distinction between flame-proofing and fire-proofing. Flame-proof material displays no after-flame or after-glow. Fire-proof material withstands I exposure to fl!lme or high temperature and still performs the function for which originally intended.

Nylon and dacron are flame-proofed by coating with plastisol and an adhesive bonding agent which will not drop off in molten drops when flame is applied. Three coats are provided to each side of the fabric. As. a result the burning stops when flame is recovered. Flame-proofing needs a large amount of material to be applied on cloth hence provides the fabric a different feeling.

By creating a finish that smoothens a flame as a fire-extinguisher does fire-proofing is done. Carbon tetra chloride or carbon tetra oxide has this effect. They cut off the oxygen supply essential for burning. Another principle is to heat the fabric with chemicals, which when comes in contact of fire, melt and cover the fabric with non-inflammable film.

With carbonates and ammonium salts results treatment in the creation of non-combustible vapours. At home, fabric can be made mildly fire retardant by dipping the dry cloth in solution of 30 percent poric acid and 70 percent borax. The fire resistant finish does not change the appearance of fabric but they should not be ironed with hot iron.

7. ***Anti-bacterial finishes***: On the fabrics germicides can be applied which are colourless and odourless, in order not to stain the fabric. It should not effect the dyes or finishes of the fabric, should not irritate the skin and should not be removed by first few washings.

8. ***Wash and Wear finish***: The fabric which is to be washed, dried and worn without ironing is given this finish. There are three kinds of methods of making the fabric wash and wear-

(i) by utilising non-resin treatment i.e. the cotton cellulose is changed by cross-linking-the forming of the bridge between two cellulose chains to produce changes in physical properties of cotton. The weakness of this method is dearth of fabric-durability in laundering.

(ii) by finishing the fabric with different kinds of resins, to make the fabric resistant to dry and wet creasing. and

(iii) by using hydrophobic fibres, e.g. nylon and polyester, that do not absorb water.,

9. ***Water-proofing***: It is intended for the fabrics which are to be water- repellent, or spot resistant. i.e. umbrellas, raincoats. Water-proofmg closes the pores of the fabric. The methods are the treatment with-

(i) Bituminous materials like tea.

(ii) Linseed oil or other drying oils.

(iii) Combinations of method I, 2, 3 and 4.

(iv) Insoluble metallic compound such as aluminium soap, fundamental acetate of aluminium, cupramonium.

(v) Paraffm or mixed waxes.

10. ***Water-repellentfinish*** : These do not close the pores of the fabric against air hence they are not so warm to wear. The fabrics are treated to lessen their affinity for water.

The methods utilised are—

(i) Application of paraffm wax and metallic salts, usually, aluminium acetate.

(ii) Application of organic solvent solution of waxy material and metallic salts.

(iii) Application of cation active compounds derived from ammonium hydroxide. These combine with fibre. The cloth feels soft and less oily to touch.

(iv) A combination of water repellency to wrinkle resistance with silicon alloy treatment.

(v) Use of a resin based material.

The fabrics after treatment vary in the degree of resistance from spot rain and shower .

11. ***Sanforising (Shrinkage control)*** : All fabrics after few washes tend to shrink, and become a bit thicker. To inhibit this the length of the ready fabric is contracted by passing the wet fabric between the layers of sanforising thick cloth and heated metallic cylinders.

12. ***Plastic coating***: The word plastic implies something that can be shaped or moulded. Plastic coated fabrics differ from plastic sheets in a way that the former has woven or knitted cloth to support it. The fabrics may be supported-

(i) having applied plastic substance to the cloth, passing the fabric under a knife to scrape off excess coating, baking in oven and ironing.

(ii) by clipping the fabric in plastic mate:rial.

(iii) by pressing a sheet of plastic against a cloth.

(iv) by melting the plastic with heat so that it adheres to cloth permanently. It is also called laminating.

Dyeing

Colour is extremely significant in textiles. Unti11865, the dyes and pigments utilised for colouring were obtained from plants, animals and insects. The dye was first invented by Pertin. Germany became the foremost centre for synthetic dyes.

Dyes

Dyes are the subtle particles, thoroughly soluble in water or some other carrier to penetrate the fibre. They are fixed by chemical action, heat or other treatment. Undissolved particles stay on the outside and then the colours have poor fastness to crocking and bleeding.

Pigments: These are the synthetic resins in an organic solvent or are insoluble colour particles that are held on the surface of a fabric by a binding agent because they have poor affmity for fibres. Their use is quick, simple and economical. Any colour can be utilised on any fibre since the pigment is held mechanically. They are also mixed with spinning solutions for man- made fibre. Stiffening, crocking and fading are some of the problems encountered with pigments but proper application gives excellence in light fastness and washability.

Process of dyeing: Dyeing process is highly complicated because fibres have an affinity for particular dyes and not for others. The thermoplastic fibres are not easy to dye because their absorbency is low. In order, that the fabric is coloured, the dye must penetrate the fibre and either combine chemically with it or be locked inside the fibre,

Dyeing process comprises—

1. Treatments to remove the media.
2. Applying dye in acid, alkaline or other media.
3. Wetting the fibres.

The dye reacts with the surface molecules first. Moisture and heat swells the fibres. The molecule chains further move separate, more reactive groups are exposed to dye. During drying the chains move back together trapping the dye in the fibres.

In wool the salt linkages provide the fibre its functional properties. The acid in the bath activates the salt linkages and becomes fixed at the amino sites. The dye reacts and displaces the acid and gets adhered to the carboxyl group. Nylon has fewer dye sites--0ne amino group for every 30 in wool.

Cellulosic fibres are absorbent due to the many hydroxyl groups but most dyes do not combine chemically with them. The acrylic fibres have anionic dye sites that react with cationic dye stuffs. They have low moisture absorption. But when heat is applied the kinetic motion of the polymer chains creates space for dye to enter. Polyester and olefms are the most problematic to dye since they have no dye sites. The dyes with carrier are forced inside the fibre through the heat which causes the fibre to swell.

Dyeing is done by comprising even application of dye by hot water, steam, and dry heat. Accelerators and regulators are used to regulate penetration of the dye. The dye can be applied at various stages. but it has nothing to do with fastness, but has great deal to do with dye penetration. There are so many of methods utilised for dyeing the fabrics at diverse stages—

1. Dyeing before yarn spinning :

(a) Solution-dyeing : Pigments or dyes are added to the spinning solution. Such fibre is spun as it is coloured.

(b) Stock of fibre-dyeing : Dye is added to loose fibres before yam spinning. Good penetration is releived but it is rather expensive.

(c) Top-dyeing : The loose ropes of wool (tops) from the combing machine are wound into the balls placed on perforated spindles and enclosed in a tank. Dye is pumped back and forth through the wool. regular process on loose fibre and wool tops are utilised utilising a pad steam technique.

2. Yarn dyed after spinning : Yarns are dyed in skeins or packages comprising one or two points of yam wound on perforated spools or tubes. Yam dyeing is less costly than raw stock dyeing.

3. Piece-dyeing after knitting or weaving : It yields solid colour fabrics. It costs less and dying can be delayed so that fashion trends can be followed more closely.

(a) *Jig-dying* : It comprises stationary dye bath with two rollers above. The cloth is rolled back and forth every 20 minutes or so. Acetate, rayon and nylon are zig dyed.

(b) *Pad-dying* : Fabric is run through the dye bath and then between squeeze rollers which forces the dye into the fabric. The cloth runs 30-300 yards a minute.

(c) *Winch reel or beck-dyeing* : Fabric is a loose rope sewed together at the ends is lifted in and out the dye bath by a reel. It is utilised on light weight fabrics.

(d) Incessant machines called ranges are utilised for larger lots of good. They wet out, dye, wash and rise the fabric.

(e) *Cross,-dyeing :* Fabrics made of fibres from various groups are dyed They generally have two colours in the fabrics

(f) *Union-dyeing :* They are used on fibres made of different groups The finished product IS solid colour

(g) *High-temperoture-dyeing :* It is essential for good penetration on thermoplastic fibres--

(i) *Molten-metal process:* Vat colour is used The fabric is passed through U-shaped tank in which molten metal alloy is maintained at 200"F to 250°F As the cloth passes through molten metal dye is reduced under pressure. The metal is rinsed and the cloth is oxidized, scoured and rinsed

(j) *Thermocol process:* The fabrics are dyed at common temperature, dried and then heat set for 30 seconds to one minute at 350°F Colours are fast to machine wash

Printing

Printing is a thousands of years old convention in India in the beginning the dyed or undyed fabric was painted with brush or printed with wooden blocks which are still in practice Later different methods brought tremendous practice change in printing industry. The dyes utilise for printing is the same as used for dyeing, but in printing they are used in the form of pastes

Printed fabrics are generally printed off grain In all over designs this is not potent but in large checks and plaids or crosswise lines matching at searn-Iine it is not possible and slanting lines across the fabric are not desirable The reason for off grain print is that the mechanism for moving the fabric does not work properly. It can be corrected at the mill if the consumer refuses to buy off grain prints, better prints may be available in the market

I. Direct Printing

1. Lacquer printing: It is just like block-printing barring that coloured enamel or opaque ink is utilised.

2. Screen printing: It is done for 500-5000 yards commercially. The cloth is pinned on the long flat table which is covered with

thick wool felt pad. An oil cloth or cotton cloth that protects the pad takes up colour from the fabric.

The screen is prepored from of wooden or metal frames over which silk, nylon or rayon'gauze is streched. It is usually 4 ft. by 5 ft. in size. The screen is covered with sensitized gelatin film and plate of original design is reproduced on film. All but design is painted out with caustic resistant paint and unpainted gelatin is washed in warm water. A special screen is prepared for each colour in the design. A metal straight edge at the side of the table makes it easy to properly set the screen. The colour is poured on the screen and is forced through the screen by moving a squeegee on the screen. Then another screen carrying a diverse colour and outline is used.

3. Block printing: It is hand process of decorating the fabric. It is expensive and slow. A design is carved on a linoleum or wood block, with 114 inch in thickness The blocks are dipped in thin panofdye and stamped on the fabric. It is allowed to dry. Then the block carrying different colours is stamped over it to form multi-coloured patterns Uniform pressure is applied

4. Stencil printing: It has more depth and texture than other prints. A hollow eopper roller is utilised which is perforated in the shape of design. The printing paste flows through the perforations, from inside the trough, on the fabric As soon as the material leaves the cylinder rayon flock is dusted on it.

5. Warp printing: It is done on warp yarns before weaving. It gives soft and hazy patterns than other prints. Warp yarns are printed while filling yarns are white or solid colour. It is generally done on taffeta, satin rubbers, upholstery and draperies.

6. Duplex printing: It is roller-printing that prints on both the side of the fabric with same or diverse patterns. It is generally used on bed covers,

7. Direct roller printing: A cast iron cylinder roller is used around which the cloth is drawn as it is printed. The copper printed roller is carved with the design. As several rollers as are the colours are utilised. Furnisher rollers, are covered with hand rubber or brushes made of nylon, or orlan brushes. They revolve in small colour trough, pick up the colour and deposit on colour cylinders. A

blade scrapes off extra colour so that only the engraved portion of the copper roller are filled with dye.

The cloth to be printed, a rubberized blanket and black gray cloth passes between the cylinder and engraved roller. Blanket gives good surface for the sharp printing and the black cloth protects the blanket and absorbs excess colour. After printing the cloth is dried, steamed or heated with resins to set the dye.

II. Discharge Printing

It is generally done on dark back grounds. Fabric is piece dyed in any of the general methods. A discharge paste which contains chemicals to remove the colour is then printed on the fabric. Dyes that are not harmed by discharging material can be mixed with discharging paste if the colour is desired in discharge areas. Fabric is steamed to develop the design. Better dye penetration is received by piece dyeing than printing. It is difficult to get good dark colours barring by piece dyeing. Discharge prints can be detected by looking at the back side of the fabric. In the design area colour that is not removed can be seen on the background colours. Particularly around the edge of the design. Background colours can be removed by strong alkali.

III. Resist Printing

Resist printing is performed by protecting areas of the fabric which are not to be dyed by covering them with wax, waxed thread or starch or resin. The method can be used only with the dyes that cannot be discharged—

(a) Tie-Dye, (b) Batik.

(a) Tie-Dye: It was developed in Indonesia. A puff of cloth is picked up and tied below with waxed thread. This thread will inhibit the penetration of dye at this tied portion. The material is then dipped in a dye bath, taken out and dried.

The waxed thread is removed divulging an undyed area. For the use of diverse shades, fabric is generally dyed in the lightest shade first and then in the darker ones. In this way very intrecate designs can be made on fabric.

Dots, stripes, length-develped, width-wise or cross-wise can be made this way, named as bundi, laharia etc.

(b) Batik: This was originated in Java and has been perfected by the people of Indonesia. The resistant material in this case is wax. The material to be printed is first properly washed to remove starch and sizing. The cloth is then oiled well with a view to produce a smooth surface for drawing. It is again washed to remove any impurities during oiling. The cloth is dried and now the design to be printed is edged or marked on it.

For instance : If red and blue and white are desired in the design, a white cloth is taken and the areas of white and red are waxed with a cup having a spout known as "Tjanting".

The wax that is painted on has to be molten. The fabric is then imersed in blue dye bath and dried before which a fixture is utilised to make dye permanent. The fabric is boiled to eliminate the wax, which floats on the top and can be removed to be reused. After dying the area of white and blue, the fabric is waxed. The fabric is then immeersed in red dye bath and then dried. The fabric is once again boiled to remove the wax and is dried in shade.

Sometimes fixing solutions are utilised to prevent fabric from bleeding and fading.

15

Starching and Ironing of Clothes

For preparing starch have four tablespoonfuls of starch, and wet it in about as much cold water, rubbing it till all lumps are removed, and then add half a teacupful of cold water. Take ready a quart of boiling water, and into it gradually pour the starch, stirring the water briskly. Some prefer pouring the boiling water into the starch. Boil it for half an hour, and just before taking it up, stir in a bit of spermaceti, as large as a hazolnut. This prevents the starch from sticking to the irons. Loaf sugar, or salt, thrown in, yield a likewise effect, but are not rathar so good. Strain the starch through the starch-bag, and then add a very lit lIe blueing. If the starch is to be made thinner, hot water should be utilised. For dark calicos, starch, thinned with coffee, prevents the whitish appearance of common starch .

Glue Starch

Glue stiffening is generally better for linen, calicos, and some other articles, than common starch. It is prepared thus: Put a piece of glue, six inches square, into four quarts of water, having first broken it up. Boil it, and then keep it corked in a jug, to utilise on washing-days. It must be mixed with the rinsingwater. According to the wish of the washer. Millinet, and any other article can be prepared stiffer, by adding glue to common starch. To prepare glue for cementing furniture, put it in a tin cup with very little water, and set the cup in boiling water until the glue is melted.

To Prepare Gum Arabic

Put one quart of hot water to four out tees of clean gum Arabic. Cork it in a bottle, and set it away for use. It is very good, when doing up lawns, book muslins, and laces. It must be blended with the starch, at the discretion of the use.

To Prepare Beef's-Gall, or Ox-Gall

Send a bottle or jug to the bultcher, and request that it may be filled with Beef's.gall. Perfume it with any strong essence that is agreeable. Keep it corked, and in a cool place. If eventually it smells disagreeably, the smell will be removed by drying the articles in the fresh air .

Directions for Starching Thick Clothes

In starching wristbands, collars, and the like, be careful that the starch is strained, so as to be free from lumps. If the starch is thinned, after it gets cold, the whole should be run through the starch-bag, to make it even. Dip the articles, and squeeze them till the starch is evenly dissipated. They should be starched just before hanging out.

Directions for CJear Starching

A number of ladies clap their muslins till clear, and then dry them entirely. After this, they sprinkle and roll, and when thoroughly damp, pull them into shape, and iron them. This method saves some time and labour, and is thought, by those who uses it, to be the best way. Others prefer to clap and pull them till about to dry, then fold them and lay them between two plates to keep them from drying any more, and when all are thus prepared, they are ironed. Wrought muslins should incessantly be ironed on the wrong side, and plain ones on the right side. Wrought work seems much better when ironed on soft flannel.

To ash and Prepare Thread and Bobbioet Laces

Sew a white cloth on a large bottle. Wind the lace around, with much care pulling out the purl, and rolling it so as not to leave the purl and scollops outside, but covering it, as you move in rolling. Cover the lace with white muslin, and then fill the bottle with water, and put it into a kettle of cold suds made of white soap. It will be well to tie it with a string to the handle, to keep it from swaying about. Boil it for one hour, and then rinse it in fair water squeezing it with the hand. Set it in the sun, and, when perfectly dry, fold it attentively, and press it between white papers, in a large book. If any stiffening is wished, put a small gum Arabic, or a little thin starch or rice.water, in the rinsing.water. It makes the lace look more like new, to wet it with sweet oil, after it is rolled on the bottle, before it is put in to boil. The oil can be used with' a sponge or rag.

To Cleanse or Whilten Silk Lace, or Blond

Roll it on a bottle, as before directed, and set the bottle upright, in cold suds, and wet it allover with the hand and set it in the sun. Renew the suds, every day. for a week, wetting and squeezing the lace lightly. Meantime keep it in the sun, as much as feaible. Take off the lace when wet, and without rinsing and pin it back and forth on a pillow or big sheet, making the edge and scollops even, and leave it, till dry, in the sun. Lay it away in loose folds. Instead of rolling on a bottle, general articles can be laid in a flat dish, in suds, and set in the sun ten days, often renewing the suds.

To Wash a White Lace Veil

Boil it. for fifteen minutes. in a strong suds of white soap. putting it in while cold. Squeeze it ill the suds, but do not rub it. Rinse in two waters, the last a little blued, and with a little gum Arabic or rice-water added. Pin it on a sheet, to dry, intensifying the scollops with care. Iron it on the wrong side, with a piece of muslin between the veil and iron. Any lace may be washed thus.

ON IRONING

Articles to be Provided for Ironing

A very easy article, for kitchen use, is a settee, or settle, made to serve also as an ironing-table. It can be made of pine, and stained of a cherry colour. The dimensions are as follows. Length, five feet and six inches; width of the seat, one foot and eight inches; height of the seat, one foot and three inches; height of sides, (or arms of the seat,) two feet and four inches, height of back, five feet and three inches. The back is made with hinges of the height of the sides or arms, so that it can be turned down, and rest on the arms, and thus become an beller ironing-table. The seat has two lids, opening into two boxes or divisions, in one of which is kept the ironing- sheets, in the other, the other ironing-apparatus. It can be put on casters, to move easily, and have cushions, stuffed with hay and covered with dark covers. It thus serves, in winter, to set before the fire, and is a comfortable seat, protecting the back from cold, apart from serving for the other purposes. The back, when down, rests on the sides, and is objective, then up, by long books behind. Instead of this, a large ironing-board is generally used, which can be put on a small table, and should be used for no other purpose. The following articls are also required in ironing. A woollen ironing-blanket, and a linen or cotton one, to put over it; ii large fire, of solid wood and charcoal, (unless stoves or furnaces are composed); a hearth, free from cinders and ashes, and a piece of sheet-iron, in front of t he fire, on which to set the irons; (this last precaution, saves many black spots from careless ironers) three or four old holders, made of woollen, and covered with old silk, as these do not take fire like cotton or linen; a ring, or iron-stand, on which to set the irons, and a piece of board to put under it, to prohibit very hot irons from scorching the sheet; a linen or cotton wiper, one for each ironer, and a piece of beeswax, to rub on the irons when smoked. There should be, at least, three irons for each person ironing: There should also be one or two large clothes-frames, on which to hang the clothes, when ironed and a small one for the smaller articles. If the house- keeper will see that all these articles are provided, that the fires are fixed properly the clothes-frames dusted, and the ironing-sheets spread smoothly and pinned well at each corner , she will be much more sure of realising good ironing, than if she does not.

The following articles will also be found very easly. A bosom-board, for ironing shirt-bosoms. This is made, by covering, with flannel, a board, one foot and six inches long, and one foot wide. It should have two over-cases, made to fit it, like pillow-cases, to be changed and washed. A skirtboard. This is a board five feet long, two feet wide at one end, tapering to one foot and four inches wide at the other . This, also, should be covered with flannel, and have two over- cases give. It save, much trouble, in ironing the skirts of dresses, and enables. the ironer to do them faster and better. The large end ghould be put on the table, and the small end on a chair. Both these boards are convenient, when ironing muslins, or when wishing to press some little articles in tbe parlor or chamber. A uting-iron, known as also a patent Italian iron, is very easy in ironing ruffles. It is made witb a tapering hollow tube, fixed to a stand, and provided witb two beaters, to run into tbe tube. Tbe ruffles are taken, wben damp, and drawn over this tube. When persons have practised awhile, they can iron very rapidly, and very beautifully, in this mode. And the lose is trifling. Another article, called a crimping-iron, is equally. It is made of two grooved brash rollers, which, wben tiJrned, work into each other. There is some danger of cutting muslins, in crimping them. To prohibit this, the iron must be screwed to the right degree of tightness, and this can be foundout only by trying it on old muslins. When it is adjusted right, care should be taken to have it remain so: They are more costly than the fluting-irons. They cost four or five dollars in New York or Boston.

On Sprinkling, Folding, and Ironing

Wipe the dust from the ironing-board. Take the clearest water , (warm water penetrates soonest,) and laying the articles one by one on the board, sprinkle them all, laying them in different piles, one of coloured, one of flannels, one of common, and one of fine, articles. Shake, fold, and roll the fine things, wrapping them in a towel the shake and fold the rest, turning them all right side outward.

The coloured clothes ought not to lie long wet or damp, as this injures colours. The sheets and table-linen must be stretched, shaken, and folded, by two persons. The coloured clothes should be separated from the rest, by a towel. Iron all lace and needle-work on the wrong side. Iron calicos with an iron not very hot, (as a hot iron will make them fade,) and often on the wrong side. In ironing frocks, first do the waist, then the sleeves, then the skirt. Keep the skirt rolled, while ironing the waist and sleeves. Set a chair to hold the waist and sleeves while ironing the skirt. In ironing a shirt, first do the back, then the sleeves, then the collar and bosom, then the front. Laces and muslins should be laid in a basket, and carried away as Soon as dry.

Iron silk on the wrong side, when it is rather damp. The iron must not be very hot. Pink. blue, green, and yellow, change colour in ironing. Black and other dark colours are less affected.

Velvet can be ironed, only by turning up the face of an iron, and after dampening the velvet, on the wrong side, draw it over the iron.

In suds, black silk should be washed with ox-gall, and stiffened subtle with glue or gum Arabic. It looks better to wash it, by sponging it, on both sides on a table, and not squeezing or wringing it. It can be stiffened, by using the gum or glue on the wrong side, when a little damp. Sleeves changed by perspiration, and yellow Spots, can be restored, by sponging with spirits of hartshorn mixed with an equal amount of water.

Bibliography

1. Roy, Nilima, Art of Manipur, Agam Kala Prakashan, New Delhi, 1979.

   This book gives a detailed description of the art of weaving of the people of Manipur. It not only talks about the technique of production but also about the weaving tools and implements and the designs and embroidery used in the weaving process. It also talks about the future of the weaving industry of manipur with the changing times and urbanization. The weaving process is made clearer with a number of illustrations.

2. Ganguli, Milada, Naga Art, New Delhi, Oxford & IBH

   This book relates the art of weaving of the Nagas with their economic and social life in a brilliant manner. With the beautiful and numerous illustrations the variations in the weaving patterns and the dresses of the different communities and tribes of Nagaland is made clearer. This book has numerous illustrations.

3. The Arts and Crafts of Nagaland, Naga institute of Culture, Government of Nagaland, Kohima, 1968.

This book describes and illustrates the technical processes involved in the working of the Naga crafts and certain aspects of the art of Nagaland. This book deals with the spinning and the weaving process, including the dyeing and the painting of the cloth. It also talks about the designs and symbols used in the Naga textiles, which is a major characteristic of their position in the society. This book has good number of illustrations.

4. Bhavnani, Enakshi, Folk and Tribal designs of India, Taraporevala, Bombay, 1974.

   This book has tried to describe some of the rarely seen and characteristic designs and motifs created by the tribes all over India. This gives an overall picture of the crafts of North-East India, and deals with the weaving crafts in detail. It helps us to have a clearer picture of the weaving processes of the people of Assam in particular.

5. Barpujari, H.K. The comprehensive History of Assam. Vol V, Publication board of Assam, 1993.

   This book talks in detail about the different varieties of Silk and cotton used by the weaving community of Assam and from where do they get they procure their raw material. It also speaks in detail about the economic side of the weaving industry i.e. how much the government and the local people earn through the industry. This book however lacks illustrations.

6. Nyori, Tai, History and Culture of the Adis, Omsons Publications, 1993.

   This book describes a very interesting legend as to how waving was started among the Adis. It also narrates a song which describes how the cotton is grown, plucked, spun and how it is woven with cotton yarn in the loom. This book however lacks illustrations.

7. Panchani, Chandra Sheikhar, Arunachal Pradesh: Religion, Culture & society, Konark Publishers, New Delhi, 1989.

This book gives a brief outline of the weaving industry in Arunachal Pradesh. It not only speaks about the weaving technique but also speaks about the difference in designs and symbols depending on the difference in community and tribe. This book however lacks illustrations.

8. Panchani, Chandra Sheikhar, Manipur: Religion, Culture & society, Konark Publishers, New Delhi, 1987.

   This book gives a brief account of the different weaving pattern amongst the Naga tribes of Manipur. It also gives a reference of the different types of cloth woven for different communities and tries to explain as to why there is a difference in the weaving patterns among the different tribes.

9. Chattopadhyay, Kamaladevi, The glory of Indian handicrafts, Indian book Company, Delhi, 1976.

   This book gives a brief outline of the weaving techniques in manipur and Assam. It is a good book if one wishes to compare the weaving techniques and patterns in Northeast with the weaning methods used in rest of the Indian states. This book however lacks illustrations.

10. Saraf, D.N. Indian Crafts- Development and Potential, Vikas Publishing House, Delhi, 1982.

    This book gives a brief description of the handloom and weaving industry of all the seven states of Northeast India. This can be a good reference for those who want to have an overall view of the weaving technology of Northeast India. This book gives a good comparison of the weaving industry of the whole of India.

11. Saraf, D.N. In the journey of Crafts development, (1941- 1991), Sampark Publishing house, Delhi, 1992.

    This book traces the development of the craft of weaving from 1941 until 1991. It shows the different changes that have come about in the craft with time and its impact on the people and the reasons for this change. The book however lacks illustrations.

12. Elwin Verrier, The Art of Northeast Frontier of India, Northeast frontier agency, Shillong, 1959.

This book talks about the art of weaving in all the states of northeast India with beautiful illustrations.

13. Hussain, Majid, Encyclopedia of India: Assam, Rima Publications, 1994.

This book speaks in detail about the handloom industry in Assam and it's bearing on the economy. It gives the facts related to the industry; it deals with the various co-operatives and the boards that were set up to uplift the handloom industry of Assam. This book has a few illustrations.

14. Hussain, Majid, Encyclopedia of Northeast India: Arunachal Pradesh, Vol. I, Rima Publications, 1994.

This book talks in detail about the textile industry of Arunachal Pradesh. It refers to the difference in the material, texture, colours and designs among the different tribes. It also talks about the significance of a few important designs or patterns made on the cloth.

15. Hussain, Majid, Encyclopedia of India: Nagaland, Rima Publications, 1994.

It emphasizes on the spinning and the weaving method or technique used by the people of Nagaland. It also refers to the dyeing techniques and the painting of the cloth i.e. how the pigment is prepared for dyeing and much more. The book however does not have illustrations.

16. Crill, Rosemary, 'Vaishnavite Silks: The figured textiles of Assam' in The woven silks of Assam, Marg, Publication, 1998.

A detailed and a well illustrated essay, which talks about the Vaishnavite cult prevalent in Assam and its significance in the weaning industry. It also talks about the other figures, designs and themes depicted on the silk textiles of Assam. The article is well illustrated.

17. Dhamija Jasleen, 'The Survey of Embroidery Traditions' in Textiles and Embroideries of India, Mark Publications, Bombay, 1965.

This article can form a base and can give light reading on the different embroidery techniques and patterns used with beautiful illustrations, with reference to states of north east India

18. Iyer D.S.V. 'Looms', in Textiles and Embroideries of India, Mark Publications, Bombay, 1965.

This article gives a general view of the different types of looms present in India with good illustrations.

19. Mudaliar, M.P.Nachimuthu, 'Manipur, Naga and Assam Fabrics', in Textiles and Embroideries of India, Mark Publications, Bombay, 1965.

This article gives a clear picture of the textiles of manipur with reference to some legends and it discusses the different designs or patterns woven on different types of cloth.

20. Hussain, Majid, N.S.Olaniya, Encyclopedia of India: Arunachal Pradesh & Mizoram, Rima Publications, 1994.

This book talks about the textile industry of Mizoram and Arunachal Pradesh. It talks about the symbolic meaning of the colours and designs used in the textiles, the kind of cloth used, and the social significance of this craft for the people of Mizoram and Arunachal Pradesh.

21. Ali, A. Yusuf, A monograph of Silk Fabrics, Allahbad, 1900.

An excellent book to have a general knowledge about the various terms related to weaving. It deals with the different kinds of looms and the different types of silk and how silk is prepared. It is a well-illustrated book.

22. Indian textiles in the 20th century: Crisis in transformation, in 2000: Reflections on the arts of India, Marg publications, 2000.

This article talks about the transformation in the weaving and the textile industry in the 20th century, the reasons for this transformation, and the impact of this transformation. It talks about the weaving industry of northeast also. Te article is supported by few illustrations.

23. Barve, V.R. Complete textile encyclopedia, Russi Lal Publishers, Bombay, 1967.

    This book deals with each term related with textile, right from the sowing of cotton to the production of the cloth in the final form. The same goes for silk and other materials. Some of the meanings are also explained through line drawings.

24. Naik, D. Shailaja, Folk embroidery and traditional Handloom weaving

    This book talks in general about the terms related to weaving. This book can be used as base book to have general idea of the methods of weaving, the weaving tools and implements and other terms related to weaving. This book has a few line drawings and a few illustrations to help make the text clear.

25. Handlooms of India, Weavers Service Center.

    Through this book, an attempt has been made for the first time to indicate the different types of handlooms that exist in this vast country. There are a few line drawings and photographs, which support the written material.

26. Baldinger, Annemarie, Seiler, Classification of Textile Techniques, Calico Museum of Textiles, Ahemdabad, 1979.

    This book gives a detailed look at the basic terms related to weaving i.e. different types of looms, techniques of thread production, and techniques of Fabric production.

27. Askari Nasreen, Liz Arthur, Uncut Cloth, Merrell Holbert Publishers, London, 1999.

    This book talks about the Saris and shawls produced in the northeastern region especially in Assam and Tripura with excellent illustrations. It also talks bout the other states of India giving a good comparative study.

28. Philipa Scott, The book Of Silk, Thames and Hudson ltd, 1993.

It talks about the origin of the silk. In the Indian variety of silk, it talks about the Assamese variety of silk with some rare pieces kept in the collections of certain foreign museums. Apart from this, it gives a list of the museum world over which hold any piece of Assamese silk in their collection.

Index

A

Abrasion 18
Adipic 72
Alpaca 11
Amorphous 57
Angora 11
Applique 35
Avian 11

B

Bombyx 56
Brocades 27

C

Conveyor 19
Cozies 118
Crewel 35

D

Dampness 47
Delustred 76
Devasted 52

E

Exquisite 110

F

Filature 53
Fuzz 82

G

Georgette 56

H

Hydrolyzing 53

I

Incontinence 133

K

Keratin 57

L

Linters 63
Lustre 18

M

Martian 26
Merino 83
Mohair 11
Moths 62
Mulberry 56
Muslin 27

O

Organza, 56

P

Planters 119

R

Reeling 52
Resilience 62
Resinous 10

S

Slanted 62
Stereoselective 18

T

Textile 1
Trichomes 10

U

Untangle 83

W

Worsteds 62

Z

Zalakdozi 34
Zean 63